Jianlong Zhang

Practical Adaptive Control

Jianlong Zhang

Practical Adaptive Control

Theory and Applications

VDM Verlag Dr. Müller

Imprint

Bibliographic information by the German National Library: The German National Library lists this publication at the German National Bibliography; detailed bibliographic information is available on the Internet at http://dnb.d-nb.de.

Cover image: www.purestockx.com

Publisher:
VDM Verlag Dr. Müller Aktiengesellschaft & Co. KG , Dudweiler Landstr. 125 a, 66123 Saarbrücken, Germany,
Phone +49 681 9100-698, Fax +49 681 9100-988,
Email: info@vdm-verlag.de

Zugl.: Los Angeles, University of Southern California, Ph.D. Dissertation, 2006

Produced in USA and UK by:
Lightning Source Inc., La Vergne, Tennessee, USA
Lightning Source UK Ltd., Milton Keynes, UK
BookSurge LLC, 5341 Dorchester Road, Suite 16, North Charleston, SC 29418, USA

ISBN: 978-3-639-04759-2

DEDICATION

To my beloved wife Yanyan and son Alex

ACKNOWLEDGEMENTS

I would like to express my deepest appreciation to the following people, whose endless and enthusiastic support has made this dissertation possible.

I would like to give millions of thanks to my advisor, Professor Petros A. Ioannou, who has been a great mentor and friend. Without his insightfulness, patience and support, I would never have come this far. I would also like to give my special thanks to Professor Michael G. Safonov for his precious inspirations and valuable discussions. I am also in debt to Professor Maged Dessouky, Professor Leonard Silverman, Professor Edmond A. Jonckheere and Professor Anastasios Chassiakos at CalState Long Beach for their critical comments.

I am very grateful to Dr. Alice C. Parker for providing the great opportunity of participating DARPA Grand Challenge 2005, Dr. Xiaoyun Lu, Dr. Anouck Girard and Dr. Susan Dickey for their help during my experiments at California PATH. My thanks also go to my great colleagues: Dr. Arnab Bose, Dr. Chin-I Liu, Dr. Haojian Xu, Dr. Hossein Jula, Dr. Barish Fidan, Dr. Ali Abdullah, Dr. Marios Lestas, Dr. Rengrong Wang, Dr. Ayanendu Paul, Dr. Margareta Stefanovic, Andrei Boitor, Hwan Chang, Ying Huo, Nazli Kahveci, Mattew Kuipers, Jason Levin, Yun Wang, Jake Parks and Muna Albasman, for their friendship and discussions.

The accomplishment also goes to my family for their wholehearted love.

TABLE OF CONTENTS

LIST OF TABLES

LIST OF FIGURES

ABSTRACT

An adaptive controller can be viewed as a combination of a control designed in the case of known parameters with an adaptive strategy accounting for the unknown parameters. It can be classified as identifier or non-identifier based, depending on the property of the adaptive strategy. In the past fifty years, adaptive control theory has reached a high level of maturity. However, its applications in safety sensitive systems are limited. Identifier based adaptive controllers can track plant changes and adjust the control parameters to counteract them. However, the conflict between parameter estimation and control may lead to poor transients during the learning process. Their nonlinear nature makes it difficult to check stability and performance bounds as done in the linear time invariant case. As a result in applications where safety is at stake practitioners are reluctant to close the loop with an identifier based adaptive controller. On the other hand, robust non-adaptive controllers can be designed to have acceptable transient behavior but may not be able to deal with unpredictable changes in the plant dynamics. In an effort to resolve this issue a bank of robust controllers is designed to cover the possible parametric uncertainty together with a switching logic searching the appropriate controller on line. These controllers, referred to as non-identifier based, are designed with the assumption that the pre-chosen controller bank contains at least one stabilizing controller. They may not guarantee good transients and it is not straightforward how to design the bank of controllers to cover the possible parametric uncertainty.

The objective of this dissertation is to develop robust adaptive control algorithms that can be implemented safely to achieve desired performance. As we show in the model reference control case, by combining identifier and non-identifier based adaptive schemes, the resulting scheme, referred to as *safe adaptive control*, is superior in performance and robustness than existing ones. In addition, we investigate the applications of conventional robust adaptive controllers in intelligent transportation systems. With theoretical analysis, simulations and actual implementation, we show that robust adaptive controllers can be implemented safely and achieve desired performance in real time applications.

CHAPTER 1 : INTRODUCTION

According to Webster's dictionary, to *adapt* means to "change (oneself) so that one's behavior will conform to new or changed circumstances." The words "*adaptive systems*" and "*adaptive control*" have been used as early as 1950 [4]. This generic definition of adaptive systems has been used to label approaches and techniques in a variety of areas despite the fact that the problems considered and approaches followed have very little in common. A generalized *adaptive controller* can be characterized as a parameterized controller whose control parameters are adjusted by certain adaptive strategy. It is classified as *identifier based* and *non-identifier based*, depending on whether or not the adaptive strategy contains any identification algorithm to identify the plant model on-line. An identifier-based adaptive controller contains an estimator that estimates the unknown plant parameters on-line, and the control parameters are tuned using the parameter estimates at each time point. The use of an on-line parameter estimator makes it possible to track changes in the plant parameters and accommodate their effects. A non-identifier based adaptive controller regulates its control parameters inside a pre-chosen set without using any estimate for the plant. Various non-identifier based schemes were proposed to relax the assumptions involved in the identifier based cases, buy they all take the strong assumption that at least one group of stabilizing control parameters stays in the pre-chosen set. This assumption could be violated in the case of drastic changes in the plant due to large parameter changes, component failures etc.

1

The design of autopilots for high-performance aircraft was one of the primary motivations for active research in adaptive control in the early 1950s. Aircraft operate over a wide range of speeds and altitudes, and their dynamics are nonlinear and conceptually time varying. For a given operating point, the complex aircraft dynamics can be approximated by a linear model. For example, for an operating point i, the longitudinal dynamics of an aircraft model may be described by a linear system of the form [47]:

$$\dot{x} = A_i x + B_i u, \quad x(t_0) = x_0$$
$$y = C_i^T x + D_i u \tag{1-1}$$

where the matrices A_i, B_i, C_i, D_i are functions of the operating point i, x is the state, u is the input and y is the measured output. As the aircraft goes through different flight conditions, the operating point changes leading to different values for A_i, B_i, C_i, D_i. Because the measured output carries information about the state x and parameters, one may argue that in principle, a sophisticated feedback controller could be able to learn the parameter changes, by processing the output $y(t)$, and use the appropriate adjustment to accommodate them. This argument led to a feedback control structure on which adaptive control is based. The controller structure consists of a feedback loop and a controller with adjustable gains as shown in Figure 1-1.

The way of adjusting the controller characteristics in response to changes in the plant and disturbance dynamics distinguishes one scheme from another.

2

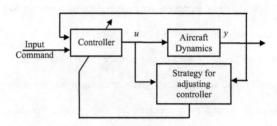

Figure 1-1: General adaptive control structure for aircraft control.

1.1 Identifier Based Adaptive Controllers

The most investigated identifier-based adaptive controller is characterized by the combination of an on-line parameter estimator, which provides estimates of unknown parameters at each instant of time, with a control law, which is that is motivated from the known parameter case. The way the parameter estimator, also referred to as adaptive law in the dissertation, is combined with the control law gives rise to two different approaches. In the first approach, referred to as indirect adaptive control, the plant parameters are estimated on-line and used to calculate the controller parameters. In other words, at each time t, the estimated plant is formed and treated as if it is the true plant in calculating the controller parameters. This approach has also been referred to as *explicit adaptive control*, because the controller design is based on an explicit plant model. In the second approach, referred to as direct adaptive control, the plant model is parameterized in terms of the desired controller parameters which are estimated directly without intermediate calculations involving plant parameter estimates. This approach has also been referred to as *implicit adaptive control* because the design is based on the estimation of an implicit plant model.

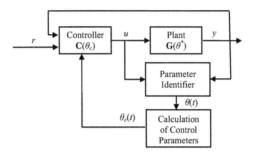

Figure 1-2: Diagram of an indirect adaptive controller.

The basic structure of indirect adaptive control is shown in Figure 1-2. The plant model $G(\theta^*)$ is parameterized with respect to some unknown parameter vector θ^*. For example, for a linear time-invariant (LTI) single-input single-output (SISO) plant model, θ^* is a vector with the unknown coefficients of the numerator and denominator of the plant model transfer function. An on-line parameter estimator generates an estimate $\theta(t)$ of θ^* at each time t by processing the plant input u and output y. The parameter estimate $\theta(t)$ specifies an estimated plant model characterized by $G(\theta(t))$, which for control design purposes is treated as the "true" plant model and is used to calculate the controller parameter or gain vector θ_c by solving a certain algebraic equation $\theta_c(t) = F(\theta(t))$ that relates the plant parameters with the controller parameters at each time t. The form of the

4

control law $C\left(\theta_c(t)\right)$ and algebraic equation $\theta_c(t) = F\left(\theta(t)\right)$ is chosen to be the same as that of the control law $C\left(\theta_c^*\right)$ and equation $\theta_c^* = F\left(\theta^*\right)$ which could be used to meet the performance requirements for the plant model $G\left(\theta^*\right)$ if θ^* was known. It is clear that with this approach, $C\left(\theta_c(t)\right)$ is designed at each time t to satisfy the performance requirements for the estimated plant model $G\left(\theta(t)\right)$ rather than the actual plant $G\left(\theta^*\right)$. Therefore, the main problem in indirect adaptive control is to choose the class of control laws $C\left(\theta_c\right)$ and the class of parameter estimators that generate $\theta(t)$, as well as the algebraic equation $\theta_c = F\left(\theta\right)$ so that $C\left(\theta_c\right)$ meets the performance requirements for the plant model $G\left(\theta^*\right)$ with unknown θ^*.

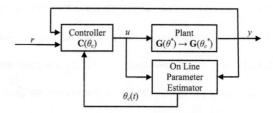

Figure 1-3: Diagram of a direct adaptive controller.

Figure 1-3 shows the structure of direct adaptive control. In this case, the plant model $G\left(\theta^*\right)$ is parameterized in terms of the unknown controller parameter vector θ_c^*, for which $C\left(\theta_c^*\right)$ meets the performance requirements, to obtain the plant model $G_c\left(\theta_c^*\right)$

5

with exactly the same input/output characteristics as $G(\theta^*)$. The on-line parameter estimator is designed based on $G_c(\theta_c^*)$ instead of $G(\theta^*)$ to provide the direct on-line estimate $\theta_c(t)$ of θ_c^* at each time t, by processing the plant input u and output y. The estimate $\theta_c(t)$ is then used in the control law without intermediate calculations. The choice of the class of control laws $C(\theta_c)$ and parameter estimators that generate $\theta_c(t)$ so that the closed loop plant meets the performance requirements is the fundamental problem in direct adaptive control. The properties of the plant model $G(\theta^*)$ are crucial in obtaining the parameterized plant model $G_c(\theta_c^*)$ that is convenient for on-line estimation. As a result, direct adaptive control is restricted to certain classes of plant models. In general, not every plant can be expressed in a parameterized form involving only the controller parameters that is suitable for on-line estimation.

In general, the ability to parameterize the plant model with respect to the desired controller parameters is what gives us the choice to use the direct adaptive control approach. Note that Figures 1-2, 1-3 can be considered having the exact same structure if in Figure 1-3 we add the calculation block $\theta_c(t) = F(\theta_c(t)) = \theta_c(t)$. This identical-in-structure interpretation is often used in the literature of adaptive control to argue that the separation of adaptive control into direct and indirect is artificial and is used simply for historical reasons. In general, direct adaptive control is applicable to SISO linear plants which are minimum-phase, as for this class of plants the parameterization of the plant with respect to the controller parameters for some controller structures is possible.

6

Indirect adaptive control can be applied to a wider class of plants with different controller structures but it suffers from the problem of loss of stabilizability problem explained as follows: As shown in Figure 1-2, the controller parameters are calculated at each time t based on the estimated plant. Such calculations are possible provided the estimated plant is controllable and observable or at least stabilizable and detectable. Since these properties cannot be guaranteed by the on-line estimator in general, the calculation of the controller parameters may not be possible at some points in time or it may lead to unacceptable large controller gains. Solutions to this stabilizability problem are possible at the expense of additional complexity. Efforts to relax the minimum-phase assumption in direct adaptive control and resolve the stabilizability problem in indirect adaptive control led to adaptive control schemes where both the controller and plant parameters are estimated on-line, leading to combined direct/indirect schemes that are usually more complex [38].

The principle behind the design of direct and indirect adaptive control shown in Figures 1-2 and 1-3 is conceptually simple. The form of the control law is the same as the one used in the case of known plant parameters. In the case of indirect adaptive control the unknown controller parameters are calculated at each time t using the estimated plant parameters generated by the on-line estimator, whereas in the direct adaptive control case the controller parameters are generated directly by the on-line estimator. In both cases the estimated parameters are treated as the true parameters for control design purposes. This design approach is called *certainty equivalence* (CE) and can be used to generate a wide class of adaptive control schemes by combining different on-line parameter estimators with different control laws. The idea behind the certainty equivalence approach is that as the parameter estimates $\theta_c(t)$ converge to the true ones θ_c^*, the performance of the adaptive controller $C(\theta_c)$ tends to that of $C(\theta_c^*)$ used in the case of known parameters. In some approaches, the control law is modified to include nonlinear terms and this approach deviates somewhat from the CE approach. The principal philosophy, however, that as the estimated parameters converge to the unknown constant parameters the control law converges to that used in the known parameter case remains the same.

In the case where the plant can be described by a member of a known class of admissible transfer functions, the set of candidate control parameters is designed such that for each admissible plant model there is a least one candidate that meets prescribed specifications [2, 19, 52, 53]. The adaptive strategy, or called *supervisor* in this case, is designed to guide the switching process among the candidates with certain properly constructed cost function, as shown in Figure 1-4. In this case, the supervisor is allowed to tune the parameterized controller in a more efficient way than the traditional robust adaptive schemes based on the pre-knowledge on the plant. Therefore, better performance is expected in the situation where the plant changes rapidly from one admissible model to another. The stability and performance bounds in these adaptive schemes, however, are difficult to check, as in the other identifier based adaptive schemes.

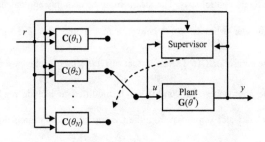

Figure 1-4: Diagram of a supervisory control system.

1.2 Non-Identifier Based Adaptive Controllers

Another class of schemes that fit the generic structure given in Figure 1-1 but do not involve on-line parameter estimators is referred to as non-identifier based adaptive control schemes. In this class of schemes, the on-line parameter estimator is replaced with search methods for finding the controller parameters in the space of possible parameters or it involves switching among candidate controllers assuming that at least one of them is stabilizing. We briefly describe the main features, advantages and limitations of these non-identifier based adaptive control schemes in the following subsections.

1.2.1 Gain Scheduling

Let us consider the aircraft model (1-1) where for each operating point i ($i=1,2,\ldots,N$) the matrices A_i, B_i, C_i, D_i are known. For each operating point i, a feedback controller with constant gains, say K_i can be designed to meet the performance requirements for the corresponding linear model. This leads to a controller, say $C(K_i)$ with a set of gains K_i ($i=1,2,\ldots,N$) covering N operating points. Once the operating point, say i, is detected the controller gains can be changed to the appropriate value of K_i obtained from the pre-computed gain set. Transitions between different operating points that lead to significant parameter changes may be handled by interpolation or by increasing the number of operating points. The two elements that are essential in implementing this approach is a look-up table to store the values of K_i and the plant measurements that correlate well with the changes in the operating points. The approach is called *gain scheduling* and is illustrated in Figure 1-5.

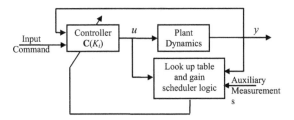

Figure 1-5: Gain scheduling structure.

The gain scheduler consists of a look-up table and the appropriate logic for detecting the operating point and choosing the corresponding value of K_i from the look-up table. With this approach, plant parameter variations can be compensated by changing the controller gains as functions of the input, output, and auxiliary measurements. The advantage of gain scheduling is that the controller gains can be changed as quickly as the

10

auxiliary measurements respond to parameter changes. Frequent and rapid changes of the controller gains, however, may lead to instability [81]. Therefore, there is a limit as to how often and how fast the controller gains can be changed. One of the disadvantages of gain scheduling is that the adjustment mechanism of the controller gains is pre-computed off-line and, therefore, provides no feedback to compensate for incorrect schedules. A careful design of the controllers at each operating point to meet certain robustness and performance measures can accommodate some uncertainties in the values of the plant parameters A_i, B_i, C_i, D_i. Large unpredictable changes in the plant parameters, however, due to failures or other effects may lead to deterioration of performance or even to complete failure. Despite its limitations, gain scheduling is a popular method for handling parameter variations in flight control [47, 75] and other systems [3].

1.2.2 Search Methods and Switching Schemes

A wide class of the non-identifier based adaptive control systems can be viewed as a supervisory control system, as show in Figure 1-4, which is composed of a set of candidate controllers and a supervisor (or *supervisory controller*) that guides the switching process among the candidate controllers without using any explicit assumptions on the plant model [18, 44, 49, 73, 84, 85]. The candidate controller set could be finite [18, 44, 84, 85], countable [49] or even uncountable [73]. With the switching process, a non-identifier based adaptive control scheme leads to a hybrid dynamic system, whose parameters change in a piece-wise constant way. Though the identifier based adaptive schemes in [17, 22, 35, 36, 52-54, 57-59] also fall into the

11

concept of supervisory control, their supervisors have certain estimators to track the changes in the plant.

All the non-identifier based adaptive control schemes were proposed to relax the assumptions used in the identifier-based adaptive schemes, but they have another strong assumption that the pre-chosen candidate controller set contains at least one stabilizing controller. The adaptive controller proposed by Martensson [44], referred to as universal controller, is constructed based on the rather weak assumption that only the order of a stabilizing linear controller needs to be known for the stabilization and regulation of the output of the unknown plant to zero. An alternative adaptive controller proposed by Fu and Barmish provides improved results of stability and performance, with additional knowledge on the set of possible plants, compactness and an a priori upper bound on the plant order [18]. In the above two schemes, the control parameters are adjusted along a pre-specified path, and on-line measurements are only used to determine the times for adjusting parameters. They are theoretically sound, but not preferred in real time applications since they usually leads to poor transient responses. In [49], the model reference adaptive control (MRAC) problem is slightly modified, and it is shown that with the only knowledge of minimum phase, the error between the plant output and the reference model output can be regulated arbitrarily small in an arbitrarily short time. Though this adaptive controller provides nice analytical results on performance, its practical application is questionable as it leads to a high gain controller and the effect of non-zero disturbances is not addressed.

An interesting non-identifier based adaptive scheme, called unfalsified control, is proposed by Safonov in [65]. In this scheme, all the candidate controllers are evaluated on-line using certain cost function constructed using the so-called *fictitious reference signals* [65], and the ones falsified by the on-line measurements are eliminated. No stability analysis has been considered, though the adaptive scheme is shown to work well in some simulations. A modified adaptive scheme with a finite number of candidate controllers has been presented in [84, 85]. In this scheme, the supervisory controller evaluates the candidate controllers without using any plant model, and hysteresis switching algorithm is adopted from [54] to preclude the possibility of unbounded chatter. As indicated in [85], model based supervisory control, such as in [52, 58], may not be able to recognize the stabilizing candidate controllers when the assumptions on the plant model fail. The supervisor designed in [84, 85], on the other hand, can always guarantee system stability as long as some plant independent assumptions are satisfied and there is at least one finite-gain stabilizing controller. The same idea has been extended to the case where the controller set is infinite in a recent study [73]. However, it is not addressed how the candidate controllers should be designed, and the performance is not clearly studied. To guarantee the feasibility of this method, all the candidate controllers have to be "stably causally left invertible" [84]. These issues restrain the non-identifier based scheme investigated in [73, 84, 85] from practical applications.

While all the above methods provide another set of tools for dealing with plants with unknown parameters they cannot replace the identifier based adaptive control schemes. Their stability properties are established with that the number of switches is

13

finite. One advantage, however, is that once the switching is over, the closed loop system is LTI and it is much easier to analyze its robustness and performance properties. This LTI nature of the closed loop system at least between switches, allow the use of the well established and powerful robust control tools for LTI systems [5] for controller design. These approaches are still at their infancy and it is not clear how they affect performance as switching may generate bad transients with adverse effects on performance.

1.3 Why Adaptive Control

The choice of adaptive control as a solution to a particular control problem involves understanding of the plant properties as well as of the performance requirements. The following simple examples illustrate situations where adaptive control is superior to linear control. Consider the scalar plant

$$\dot{x} = ax + u$$

where u is the control input, x the scalar state of the plant, and the parameter a is unknown. We like to choose the input u so that the state x is bounded and driven to zero with time. If a was a known parameter, then the linear control law

$$u = -kx, \ k > |a|$$

can meet the control objective. In fact if an upper bound $\bar{a} \geq |a|$ is known the above linear control law with $k > \bar{a}$ can also meet the control objective. On the other hand if a changes so that $a > k > 0$ then the closed loop plant will be unstable. The conclusion is that in the lack of an upper bound for the plant parameter no linear controller could stabilize the plant and drive the state to zero. On the other hand, the adaptive control law

$$u = -kx, \ \dot{k} = x^2$$

guarantees that all signals are bounded and x converges to zero no matter what the value of the parameter a is. This simple example demonstrates that adaptive control is a potential approach to use in situations where linear controllers cannot handle the parametric uncertainty.

Another example where adaptive control law may have superior properties than the traditional linear schemes is the following: Consider the same example as above but with an external bounded disturbance d:

$$\dot{x} = ax + u + d$$

The disturbance is unknown but it can be approximated as

$$d = \sum_{i=1}^{N} \theta_i^* \phi_i(x,t)$$

where $\phi_i(x,t)$ $(i=1,2,...,N$) are known functions and θ_i^* are unknown constant parameters. In this case if we use the linear control law

$$u = -kx$$

with $k > \bar{a} \geq |a|$, we can establish that x is bounded and at steady state

$$|x| \leq \frac{d_0}{k-a}$$

where d_0 is an upper bound for $|d|$. It is clear that by increasing the value of the controller gain k, we can make the steady state value of x as small as we like it. This will

15

lead to a high gain controller, however, which is undesirable especially in the presence of high frequency unmodeled dynamics. In principle, however, we cannot guarantee that x will be driven to zero for any finite control gain in the presence of non-zero disturbance d. The adaptive control approach is to estimate on-line the disturbance d and cancel its effect via feedback. The following adaptive control law can be shown to guarantee signal boundedness and convergence of the state x to zero with time

$$u = -kx - \hat{d}, \; \hat{d} = \sum_{i=1}^{N} \theta_i \phi_i (x,t), \; \dot{\theta}_i = x \phi_i (x,t)$$

where $k > \bar{a} \geq |a|$ assuming of course that $\bar{a}$ is known otherwise k has to be estimated too. Therefore, in addition to stability, adaptive control techniques could be used to improve performance in a wide variety of situations where linear techniques would fail to meet the performance characteristics.

1.4 A Brief History

The study of adaptive control began as far back as in the 1950s [4]. One of the primary motivations for research on adaptive control was the autopilot design for high-performance aircraft in the early 1950s. The dynamics of high-performance aircraft change drastically when it flies from one operating point to another, and cannot be handled by one constant-gain feedback controller. A sophisticated controller, such as an adaptive controller, was desired to counteract the drastic changes and provide satisfactory performance. Different adaptive control schemes, including adaptive pole placement

16

control (APPC) and model reference adaptive control (MRAC) were proposed to solve the problem [32, 86] . However, the lack of stability analysis and the lack of understanding of the properties of the proposed adaptive control schemes coupled with a disaster in flight test caused the interest in adaptive control to diminish.

In the 1970s, there were several breakthrough results in adaptive control. Different adaptive control schemes with well-established stability properties were introduced along with several successful applications [15, 20, 50, 60, 61]. In these adaptive control designs, the plant is described by an ideal linear time invariant (LTI) model free of modeling uncertainties, disturbances and noises. The actual plant, most likely, deviates from the plant model on which control design is based. Hence it is not surprising to find out that these adaptive laws and control schemes designed based on the ideal plant model may lead to instability when applied on the actual plant [15]. This non-robust behavior became a controversial issue in the early 1980s, and motivated many researchers to study the mechanisms of instabilities and find ways to counteract them. These studies leaded to a body of work known as *robust adaptive control* [25]. The adaptive laws proposed in the 1970s were redesigned with various robust modifications, such as dynamic normalization, leakage, dead zone and parameters projection. It has been shown that the adaptive controller designed with robust adaptive laws can guarantee system stability when the modeling errors are "reasonably" small. Furthermore, the tracking errors are bounded in the mean square sense [25]. These robust modifications are also adopted for adaptive control of linear time varying systems [81]. "The solution of the robustness problem in adaptive control led to the solution of the long-standing problem of controlling a linear plant whose parameters are unknown and changing with time" [27].

From the mid 80's, several groups of researchers started to look for alternative methods of controlling plants with unknown parameters [18, 44, 49, 65, 85]. These methods avoid the use of on-line parameter estimators in general and use search methods or switching logic to stabilize the closed-loop system. Research in these non-identifier based adaptive control techniques is still going on and issues such as robustness and performance are still to be resolved.

In recent years, considerable research efforts have been made to deal with the design of stabilizing controllers for classes of nonlinear plants. Most of the research efforts have assumed that the plant nonlinearities are known [34, 71]. In the 1990s, adaptive control has also been introduced into nonlinear systems with unknown parameters [40]. A series of adaptive control schemes were motivated by the theory of "feedback linearization" in the 1980s and "matching conditions" [35, 36, 40, 88]. The very restrictive "matching conditions" were later relaxed to "extended matching conditions", which leaded to one important ingredient in nonlinear adaptive control, called "adaptive backstepping control" [40]. However, it is still an open issue how to deal with the systems whose unknown parameters enter the parametric models in a nonlinear fashion. One suggested solution is to design the adaptive controller based on a neural network that approximates the true nonlinear plant. It is well known that a one-hidden layer neural network, when including sufficiently many neurons in the hidden layer, can approximate any continuous function defined on a compact set with an arbitrarily small error [14]. Hence we can use neural network to approximate a nonlinear system whose unknown parameters appear nonlinearly, but theses unknown parameters appear linearly in the approximate system.

1.5 Research Motivation

During the past fifty years, adaptive control theory has reached a high level of maturity. Extensive adaptive schemes have been developed with well established stability properties and many successful applications of adaptive control have been reported. However, applications of adaptive control in safety sensitive systems, such as aircraft, are still very limited. For the non-identifier based adaptive strategies, most of them are not interested in real time applications at all since they adjust control parameters along a pre-specified sequence and usually lead to poor performance. The non-identifier based adaptive strategy considered in [73, 84, 85] is rather interesting since it activates candidate controllers (or regulates control parameters) based on certain controller related costs. The stability bounds are easy to establish in the case where all the sub-systems are LTI. However, it has not been fully addressed how the candidate controllers should be designed, and the performance is not clearly investigated. As for identifier based adaptive controllers, their most attractive feature is that they can track the plant changes and adjust the corresponding control parameters to counteract them. However, they suffer from the conflict between parameter estimation and control, and may lead to worse transient performance than a non-adaptive controller when poor initial estimates are used. Large

19

transient oscillations may be created by the adaptive controller to improve the estimation quality. Furthermore, when the aircraft flies from one point to another, its dynamics change drastically and the traditional identifier-based adaptive controller may not be able to tune the control parameters fast enough to guarantee good performance. Another drawback of the identifier-based adaptive control in addition to possible undesirable transients is that its nonlinear nature makes difficult to check stability and robustness bounds as done in the LTI case. As a result in applications where safety is at stake such as aircraft control practitioners are reluctant to close the loop with an identifier-based adaptive controller. On the other hand, the popularly used control laws in safety sensitive systems, such as robust controllers and gain-scheduling techniques, do not account for unpredictable changes in the plant and may deteriorate performance when system failure appears.

Consider the control system shown in Figure 1-6, in which the non-adaptive controller C_1 is a robust controller that can always stabilize the plant (aircraft), and C_2 is a robust identifier based adaptive controller designed based on certain reasonable but unverified assumptions. The question arises: can we design a supervisor, as shown in Figure 1-6, so that the closed loop system is guaranteed to be stable and the adaptive controller C_2 will be employed only when it will improve system performance? The research work presented in Chapters 2 and 3 is on the way to solve such a problem. The non-identifier based adaptive scheme proposed in [84, 85] is revisited in Chapter 2 and then modified in Chapter 3 to design the supervisor in Figure 1-6. The overall controller formed by the supervisor and the two candidate controllers is a generalized adaptive controller, and it is referred to as the *safe adaptive controller* in our work since the adaptive candidate controller is activated in a conservative but safe manner. It is shown that the system stability is guaranteed when the supervisor is designed properly. Simulation results indicate that the proposed safe adaptive controller can achieve perform superior to that can be achieved by either of the candidate controllers. The significance of the safe adaptive control is that it will push the existing adaptive schemes into the application fields where their stability and performance are doubted. It should be noted that the term "safe adaptive control" was also used in the study [74] for solving a different problem (see Remark 3-2 in Chapter 3).

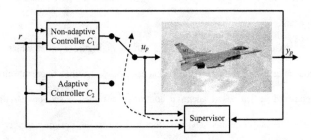

Figure 1-6: Safe switching between adaptive and non-adaptive controller.

Nevertheless, conventional identifier based adaptive controllers can also be implemented safely to achieve desired performance if designed properly, which is demonstrated using Automated Highway Systems (AHS) in this dissertation.

During the last decade, semi-automated vehicles with Adaptive Cruise Control (ACC) system have already entered the market. The ACC system enables the vehicle to automatically follow the preceding vehicle in the same lane. It has been found out the time headway used by the ACC system is a critical parameter of an AHS system. However, not many results have been reported on the design of ACC systems with variable time headway on real vehicles. In Chapter 4, we design adaptive vehicle following controller that can provide desired stability properties for a class of variable time headway including those in [12, 66, 77, 83, 91] as well as the constant time headway. The performance of the designed adaptive vehicle following controller is demonstrated by the simulations using a validated nonlinear passenger vehicle model. In Chapter 5, another vehicle following controller is proposed that provides better performance on the microscopic level with beneficial effects on fuel economy and traffic stability. This vehicle following controller is formed by a nonlinear reference speed generator for disturbance rejection and an adaptive speed tracking controller. Simulations are conducted with a validated nonlinear truck model to demonstrate the benefits. In Chapter 6, automated trucks with the adaptive controllers designed in Chapter 4 and 5 are implemented in an automated transportation system between inland and port (ACTIPOT) to improve the performance of container terminals and meeting the challenges of the future in marine transportation. All the practical applications investigated in this dissertation have indicated that the conventional robust adaptive controllers can be implemented safely and achieve desired performance with proper designs.

CHAPTER 2 : NON-IDENTIFIER BASED ADAPTIVE CONTROL SCHEME WITH GUARANTEED STABILITY

2.1 Introduction

A generalized adaptive controller can be viewed as a parameterized controller whose control parameters are adjusted by certain adaptive strategy [51]. It can be classified as *identifier based* or *non-identifier based*, depending on whether or not certain parameter estimator is employed. An identifier-based adaptive controller contains an estimator that estimates the unknown plant parameters on-line, and the control parameters are tuned using the parameter estimates at each time t. Extensive research efforts on adaptive control have been dedicated to identifier based adaptive controllers, since the use of identifiers gives them the distinguished capacity of stabilizing a plant whose unknown parameters may change slowly with time. Different identifier based adaptive control designs can be found in [27, 40] and the references insides. A non-identifier based adaptive controller has no identifier to track the changes in the plant, and the control parameters stay in a pre-chosen set, which could be finite [18, 44, 84, 85], countable [49], or even uncountable [73]. Most of the non-identifier based adaptive control schemes in literature were proposed to relax the assumptions on the plant model required in the identifier based adaptive schemes. In many cases [18, 44, 49], the control parameters are adjusted along a pre-determined path, and on-line measurements are only used to

23

determine the times for adjusting parameters. An interesting exception, proposed by Safonov [65], is called unfalsified control. In this adaptive strategy, the control parameters are tuned based on certain performance related costs generated without using any plant model, and the control parameters falsified by the on-line measurements are eliminated. It has been shown to be able to provide good performance in simulations, though no stability analysis has been considered. A modified adaptive scheme is proposed in [84, 85], in which no candidate controller (or control parameter) will be eliminated and the hysteresis switching algorithm investigated in [48, 54] is employed to supervise the switching process among a finite set of candidate controllers. It is shown that given a set of candidate controllers, the closed-loop system is guaranteed to be stable as long as the adaptive scheme is feasible and there is one "finite-gain" stabilizing candidate controller. In another study by Stefanovic et al. [73], the same idea has been extended to the case of continuously parameterized controller set. The most attractive feature of this adaptive scheme is that it can avoid the instability raised by model mis-matching since no explicit assumptions on the plant model are required [84, 85]. However, all the candidate controllers have to be "stably causally left invertible" [84] to guarantee the feasibility of the adaptive control scheme.

In this chapter, we revisit the non-identifier based adaptive scheme originally proposed by Wang et al. in [84, 85] for the general case where all the sub-systems are nonlinear, and stronger stability results are developed (Theorem 2-1). Then we focus on the special case in which all the sub-systems are linear time invariant (LTI) and propose a modified cost function so that the original non-identifier based adaptive scheme can be

24

applied with any LTI controllers when the unknown plant is LTI (Lemma 2-3). The system stability is guaranteed as long as the candidate controller set contains at least one stabilizing controller, and the presence of bounded disturbances will not destroy the system stability. The rest of this chapter is organized as follows. In section 2.2, we present the problem formulation as well as some preliminary definitions. In section 2.3, we present the non-identifier based adaptive control scheme and the stability results. The concluding remarks are given in section 2.4.

2.2 Problem Formulation

2.2.1 Preliminaries

For a real function of time, $x(t)$, we define the L_p norm as

$$\|x\|_p \triangleq \left(\int_0^\infty | x(\tau) |^p \, d\tau \right)^{1/p}$$

where $|\cdot|$ is the standard Euclidean norm, for $p \in [1, \infty)$ and say that $x \in L_p$ when $\|x\|_p$ is finite. The L_∞ norm is defined as

$$\|x\|_\infty = \operatorname*{ess\,sup}_{\tau \in \mathbb{R}_+} \left(|x(\tau)| \right)$$

where $\mathbb{R}_+ = [0, \infty)$ and by *essential supremum* we mean

$$\operatorname*{ess\,sup}_{\tau \in \mathbb{R}_+} \left(|x(\tau)| \right) = \inf \left\{ a \, | \, |x(\tau)| \leq a \text{ almost everywhere} \right\}$$

We say that $x \in L_\infty$ when $\|x\|_\infty$ is finite. We define the truncation of $x(t)$ over a time interval $I \subset \mathbb{R}_+$ as

$$x_I = \begin{cases} x(t), & \text{if } t \in I \\ 0, & \text{otherwise} \end{cases}$$

and in particular we use x_τ to denote the truncation of $x(t)$ over the time interval $[0, \tau)$.

For the functions of time that do not belong to L_p, we define the L_{pe} norm as

$$\|x_t\|_p \triangleq \left(\int_0^t |x(\tau)|^p \, d\tau \right)^{1/p}$$

for $p \in [1, \infty)$. We say that $x \in L_{pe}$ when $\|x_t\|_p$ exists for any finite t. Similarly, the $L_{\infty e}$ norm is defined as

$$\|x_t\|_\infty \triangleq \operatorname*{ess\ sup}_{0 \le \tau \le t} (|x(\tau)|)$$

Since in this chapter the stability is most interested in the L_∞ norm (or equivalently in the $L_{\infty e}$ norm), $\|\cdot\|$ is used to represent $\|\cdot\|_\infty$ unless otherwise clarified. Though the definitions and analytical results in this chapter are introduced in the L_∞ norm, they can be easily extended to any other L_p norms with proper modifications.

A continuous function $\alpha : \mathbb{R}_+ \to \mathbb{R}_+$ is said to be of class $\mathcal{K}$ if it strictly increasing and $\alpha(0) = 0$. If $\alpha \in \mathcal{K}$ is unbounded, then it is said to be of class $\mathcal{K}_\infty$. A function $\beta(s,t) : \mathbb{R}_+ \times \mathbb{R}_+ \to \mathbb{R}_+$ is said to be of class $\mathcal{KL}$, if $\beta(\cdot, t)$ is of class $\mathcal{K}$ for each fixed $t \ge 0$ and $\beta(s,t)$ decreases to 0 as $t \to \infty$ for each fixed $s \ge 0$.

We consider a general finite-dimensional MIMO nonlinear system

$$\begin{cases} \dot{x} = f(x,u) \\ y = h(x,u) \end{cases} \tag{2-1}$$

where $x \in \mathbb{R}^n$, $u \in \mathbb{R}^m$ and $y \in \mathbb{R}^l$ are the system state, input and output (n, m and l are some

positive integers), respectively, $f: \mathbb{R}^n \times \mathbb{R}^m \to \mathbb{R}^n$ and $h: \mathbb{R}^n \times \mathbb{R}^m \to \mathbb{R}^l$ are two smooth enough

functions. Here, u is piecewise smooth and it represents the input that we are interested in,

i.e. the input that can be measured or controlled. The immeasurable disturbance inputs are

implicitly included in (2-1).

Definition 2-1 (stability)

The system in (2-1) is said to be (*weakly*) *stable* if for any given initial condition $x(0) \in L_\infty$

and input $u \in L_{\infty e}$, there exist a non-negative constant c and a function $\gamma \in \mathcal{K}_\infty$ such that

$$\|y_t\| \le \gamma(\|u_t\|) + c \tag{2-2}$$

holds for all $t \in \mathbb{R}_+$. Otherwise, it is said to be *unstable*. The system in (2-1) is said to be

strongly stable if for any given initial condition $x(0) \in L_\infty$ there exist a non-negative

constant c and a function $\gamma \in \mathcal{K}_\infty$ such that (2-2) holds for all $u \in L_{\infty e}$ and $t \in \mathbb{R}_+$. In

particular, if the system in (2-1) is strongly stable and the function γ in (2-2) is a linear

function, then it is said to be *finite-gain stable*.

Remark 2-1: The different stability concepts introduced in Definition 2-1 are all consistent with the stability concept for LTI systems, provided that they are detectable in the classical sense. A very similar definition was introduced in [72], referred to as *input-to-output stability*. The system (2-1) is said to be input-to-output stable if there exist some functions $\gamma \in \mathcal{K}_\infty$ and $\beta \in \mathcal{KL}$, such that for every $x(0)$ and every u, the inequality

$$\left| y(t) \right| \leq \gamma \left(\left\| u_t \right\| \right) + \beta \left(\left| x(0) \right|, t \right)$$

holds along the corresponding solution. If we consider that the immeasurable disturbance terms are bounded and incorporated in u appearing in the above equality, then the input-to-output stability defined in [72] implies the strong stability introduced in Definition 2-1.

Remark 2-2: In the work by Wang et al. [84, 85], the stability definition is introduced in the same way with the L_{2e} norm. As pointed out in [84, 85], the basic idea roots in the early work by Willems [87].

Definition 2-2 (output-input stability) [41]

The system in (2-1) is said to be *output-input stable* if there exist a non-negative integer N, and functions $\gamma \in \mathcal{K}_\infty$ and $\beta \in \mathcal{KL}$, such that for every initial state $x(0)$ and every N times continuously differentiable input u the inequality

$$\left| \begin{pmatrix} u(t) \\ x(t) \end{pmatrix} \right| \leq \gamma \left(\left\| y_t^N \right\| \right) + \beta \left(\left| x(0) \right|, t \right) \tag{2-3}$$

28

holds for all t in the domain of the corresponding solution of (2-1), where $\mathbf{y}^N = [y^T, y^{(1)T}, \ldots$ $y^{(N)T}]^T$.

Definition 2-3 (0-detectability) [41]

We say that the system in (2-1) is 0-dectectable if there exist some functions $\gamma_1, \gamma_2, \in \mathcal{K}_\infty$ and $\beta \in \mathcal{KL}$ such that for every initial state $x(0)$ and every u the corresponding solution satisfies the inequality

$$|x(t)| \le \beta\left(|x(0)|, t\right) + \gamma_1\left(\|u_t\|\right) + \gamma_2\left(\|y_t\|\right) \qquad (2\text{-}4)$$

as long as it exists.

Remark 2-3: For stabilizable and detectable LTI systems, output-input stability is equivalent to the minimum phase property, and the minimum value for N in Definition 2-2 is the relative degree [41]. 0-detectability was studied under the name *input-output-to-state stability* and it was shown to be equivalent to the detectability property for LTI systems in [39]. For completeness of the work, this point is shown in Lemma 2-2.

2.3 Problem Description

The non-identifier based adaptive control system investigated in this chapter can be viewed as a feedback interconnection of a SISO plant Σ_P, a candidate controller set Σ_C and a supervisory controller Σ_S (non-identifier based adaptive strategy), as shown in

29

Figure 2-1, in which r is the reference signal, u_c is the controller output, u_p and y_p are the input and output of Σ_P, respectively, d_u and d_y are bounded piecewise continuous disturbance signals, y_s is the disturbed plant output, and σ is the piecewise constant switching signal generated by Σ_S.

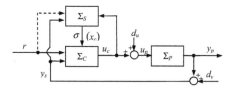

Figure 2-1: Supervisory control of a SISO plant Σ_P.

We consider the case in which Σ_P is a SISO 0-detectable system expressed as

$$\begin{cases} \dot{x}_p = f_p\left(x_p, u_p\right) \\ y_p = h_p\left(x_p, u_p\right) \end{cases} \tag{2-5}$$

where x_p, u_p and y_p are the state, input and output of Σ_P with proper dimensions, f_p is a locally Lipschitz function, and h_p is a continuous function. The candidate controller set Σ_C contains a finite number of time invariant 0-detectable controllers, C_i, $i \in \mathbf{m} = \{1, 2, ..., m\}$, where m is a finite integer. A candidate controller C_i ($i \in \mathbf{m}$) can be expressed as

$$\begin{cases} \dot{x}_{c,i} = f_{c,i}\left(x_{c,i}, y_s, r\right) \\ u_{c,i} = h_{c,i}\left(x_{c,i}, y_s, r\right) \end{cases} \tag{2-6}$$

where $x_{c,i}$ and $u_{c,i}$ are the state and output of the controller i with proper dimensions, and $f_{c,i}$ and $h_{c,i}$ are two smooth enough functions. The subscript i will be omitted when there is

30

no confusion. In particular, we are interested in the case where all the candidate controllers are of two degree-of-freedom. We denote the ith candidate controller as $C_i=\{C_{1,i}, C_{2,i}\}$, where $C_{1,i}$ and $C_{2,i}$ are the two SISO sub-controllers, as shown in Figure 2-2. The sub-controller $C_{2,i}$ can be expressed as

$$\begin{cases} \dot{x}_{c2,i} = f_{c2,i}\left(x_{c2,i}, y_s\right) \\ v_{c,i} = h_{c2,i}\left(x_{c2,i}, y_s\right) \end{cases} \tag{2-7a}$$

where $x_{c2,i}$ and $v_{c,i}$ are the state and output of $C_{2,i}$, and $f_{c2,i}$ and $h_{c2,i}$ are two smooth enough functions. $C_{1,i}$ can be expressed as

$$\begin{cases} \dot{x}_{c1,i} = f_{c1,i}\left(x_{c1,i}, e_i\right) \\ u_{c,i} = h_{c1,i}\left(x_{c1,i}, e_i\right) \end{cases} \tag{2-7b}$$

where $x_{c1,i}$ is the state of C_{1i}, $e_i = r - v_{c,i}$, and $f_{c1,i}$ and $h_{c1,i}$ are two smooth enough functions.

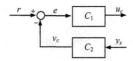

Figure 2-2: Diagram of a two degree-of-freedom controller.

The supervisory controller Σ_S, which is the adaptive strategy to be designed in this chapter, generates the switching signal σ with the on-line measurements u_c, y_s (and possible with r). $\sigma : [0, \infty) \to \mathbf{m}$ is a piecewise constant function of time, and the times at which the switching signal σ is discontinuous are called the switching times. We denote

31

$\{t_k \mid k \in \mathbf{K}\}$ as an ordered sequence of switching times. For simplicity of analysis, we assume that σ is right continuous, i.e. if t_1 and t_2 are two consecutive switching times of a feasible signal σ, then σ is constant on $[t_1, t_2)$. The supervisory controller Σ_S sets the controller state (x_c) for the active controller at each switching time.

The problem investigated in this chapter is described as the follow. For the control system given in Figure 2-1, design the non-identifier based adaptive scheme (or equivalently, the supervisory controller Σ_S in Figure 2-1) so that the closed-loop system is guaranteed to be stable whenever there is a candidate controller $C \in \Sigma_C$ that can stabilize the closed-loop system shown in Figure 2-3.

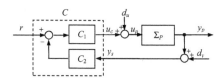

Figure 2-3: Feedback interconnection of a SISO plant Σ_P and a controller C.

Though Σ_P is SISO, the closed-loop system in Figure 2-3 is considered as a single-input multi-output (SIMO) system, whose input is r and output vector is $[y_s, u_c]^T$. Here we assume that u_c and y_s can be perfectly measured, and we use z to represent the output vector $[y_s, u_c]^T$. We denote $\mathcal{Z}$ as the set of all conceivable signals of z.

It should be noticed that many identifier based adaptive control systems, such as those studied in [52, 58], can also be presented as the system shown in Figure 2-1. The adaptive schemes in these studies, however, involve certain estimators to estimate the "distances" between the unknown plant and the admissible models. In this work, we assume that the set of candidate controllers is given, whether designed based on certain nominal models or randomly chosen. The design of supervisory controller Σ_S does not depend on any explicit assumptions for the plant except that the controller set is "consistent" (Definition 2-6 in section 2.3) with the plant and there exists at least one candidate control that can make the closed-loop system shown in Figure 2-3 finite-gain stable (Definition 2-1 in section 2.2). The idea has been investigated by Safonov et al. in different papers [65, 74, 84, 85]. We re-examine this idea for the general case where all the sub-systems are nonlinear, and stronger stability results are developed when the state is reset properly for each active controller. We also investigate the special case in which all the sub-systems are LTI. We propose a modified cost function so that the original non-identifier based adaptive scheme can be applied with any general LTI controllers when the unknown plant is LTI. The requirement that all the candidate controllers are "stably causally left invertible" [84] is removed in our adaptive scheme. The system stability is guaranteed as long as the candidate controller set contains at least one stabilizing LTI controller, and the presence of bounded disturbances will not destroy the system stability.

2.3 Adaptive Control Scheme and Stability Results

Several key concepts are introduced before we present the adaptive control scheme and the related stability results.

Definition 2-4 (matching reference signal)

For a candidate controller $C \in \Sigma_C$ described by (2-6), $\tilde{r}(C, z, x_c(0))$ is a *matching reference signal* for C if it is an element of

$$\{ r \mid C(r, y_s, x_c(0)) = u_c, z = [u_c, y_s]^T \}$$

where $C(r, y_s, x_c(0))$ represents the output of C with the input $[r, y_s]^T$ and the initial control state $x_c(0)$. $\tilde{r}(C, z, x_c(0), t)$ denotes the matching reference signal at time t.

Remark 2-4: The matching reference signal was originally defined as *fictitious reference signal* in [65]. It was also referred to as *virtual reference* in [1, 55]. Here we use the name "matching reference signal" since $\tilde{r}(C, z, x_c(0))$ in fact is the reference signal matches the data $z = [u_c, y_s]^T$ and the initial condition $x_c(0)$ in the system defined by (2-6). In this work, we use $\tilde{r}(C_i, z)$ or $\tilde{r}_i$ to represent $\tilde{r}(C, z, x_c(0))$. We will drop the initial condition in $\tilde{r}(C, z, x_c(0), t)$ when there is no confusion.

Definition 2-5 (matching controller state)

For a candidate controller $C \in \Sigma_C$, its *matching controller state* at time t is the state that C should process at time t if C had been the active controller and $\tilde{r}(C, z)$ was the reference signal.

Definition 2-6 (consistence)

The controller set Σ_C is said to be *consistent* with the plant Σ_P (for the control system shown in Figure 2-1), if $\forall C \in \Sigma_C$ has been the active controller, any controller in Σ_C processes one and only one matching reference signal.

Remark 2-5: The consistence between Σ_C and Σ_P guarantees that the existence and uniqueness of the matching reference signals, and it has to be satisfied for the feasibility of the proposed adaptive scheme. In the following analysis, it is always assumed that Σ_C is consistent with Σ_P. It is easy to verify that the consistence is guaranteed if all the candidate controllers are LTI, though the matching reference signals may contain impulses.

Definition 2-7 [14] (cost function)

For each $C \in \Sigma_C$, its associated cost function $V(C, z, t)$ is a mapping from $\Sigma_C \times \mathcal{Z} \times \mathbb{R}_+$ to $\mathbb{R}_+$,

where $\mathcal{Z}$ is the set of all conceivable signals z.

The adaptive scheme, or equivalently, the supervisory controller considered in this work is composed of three subsystems:

1. *Matching reference signal generator*: A dynamic system whose inputs are the on-line data u_c and y_s, and whose outputs are the matching reference signals $\tilde{r}_i$, $i \in \mathbf{m}$.

35

2. *Controller cost generator*: a dynamic system whose inputs are the on-line data u_c and y_s and the matching reference signals $\tilde{r}_i$, $i \in \mathbf{m}$, and whose outputs V_i (cost for C_i), $i \in \mathbf{m}$ are suitable defined controller costs.

3. *Switching logic*: a dynamic system whose inputs are the control costs V_i, $i \in \mathbf{m}$ and whose output is a piecewise constant switching signal σ, taking values in $\mathbf{m}$, which is used to define the control law $u = u_\sigma$.

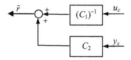

Figure 2-4: Matching reference signal generator for a controller of two degree-of-freedom.

A matching reference signal generator for a two degree-of-freedom controller is presented in Figure 2-4. Suppose Σ_C is consistent with Σ_P, and then it follows Definitions 5 and 6 that if a controller $C \in \Sigma_C$ is active during the time interval I and it is switching on with the matching controller state, then $\tilde{r}(C,z)_I = r_I$. If a controller is not assigned with the matching controller state when becoming active, then its matching reference signal may not be the same as the true reference signal on its active time interval. We say the matching reference signal generator for $C \in \Sigma_C$ is *stable with respect to* (w.r.t.) *initial state* if there exist a function such that $\gamma \in \mathcal{K}_\infty$

$$\left\| \tilde{r}\left(C,z,x_c\left(0\right)\right) - \tilde{r}\left(C,z,\tilde{x}_c\left(0\right)\right) \right\| \le \gamma\left(\left|x_c\left(0\right) - \tilde{x}_c\left(0\right)\right|\right)$$

for all $z \in \mathbf{Z}$ and $x_c(0), \tilde{x}_c\left(0\right) \in L_\infty$.

The cost function for a candidate controller $C \in \Sigma_C$ is designed as

$$V\left(C,z,t\right) = \sup_{\tau \in [0,t)} \frac{\left\| z_\tau \right\|}{\left\| \left(\tilde{r}\left(C,z\right)\right)_\tau \right\| + c} \tag{2-8}$$

where c is a positive design constant. Since all the conceivable reference signals are piecewise smooth and the disturbances are piecewise continuous, it is easy to show that z is piecewise continuous, and then the cost function in (2-8) is piecewise continuous for any $C \in \Sigma_C$. Furthermore, it is a non-decreasing function of t. A similar cost function was defined with the L_{2e} norm in [84, 85].

Remark 2-6: There are many other ways to choose the controller cost function, and the system stability can still be established. For example, it can be chosen as a smooth coercive function of $\sup_{\tau \in [0,t)} \dfrac{\left\| z_\tau \right\|}{\left\| \left(\tilde{r}\left(C,z\right)\right)_\tau \right\| + c}$.

For the switching logic, we adopt the hysteresis switching algorithm investigated in [54]. Let $\rho : \mathbb{R}_+^n \to \mathbf{m}$ denote the function whose value at $V = \{V_1,...,V_m\}^T \in \mathbb{R}_+^m$, denoted as $\rho(V)$, is the least integer $i \in \mathbf{m}$ for which $V_i \le V_j, j \in \mathbf{m}$. The hysteresis switching algorithm is defined recursively as [54]

37

$$\begin{cases} \sigma(t) = \phi\big(\sigma^-(t), V\big) = \begin{cases} \sigma^-(t), & \text{if } V_{\sigma^-(t)} < V_{\rho(V)} + \varepsilon \\ \rho(V), & \text{if } V_{\sigma^-(t)} \geq V_{\rho(V)} + \varepsilon \end{cases} \\ \sigma^-(0) = i_0 \end{cases} \tag{2-9}$$

where ε is a positive design parameter and i_0 is the initial condition of σ. The switching algorithm can also be viewed in Figure 2-5.

As we can see the supervisor designed in this paper is the same as that in [84, 85] except that the cost function is defined in a different norm.

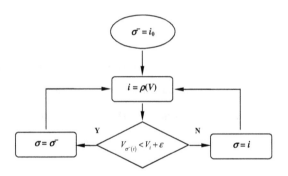

Figure 2-5: Hysteresis switching algorithm.

Lemma 2-1: If a system Σ is finite-gain stable, then $\forall x(0)$ and $c>0$, $\exists \mu \geq 0$ such that

$$\frac{\left\| (\Sigma r)_\tau \right\|}{\left\| r_\tau \right\| + c} \leq \mu \tag{2-10}$$

for all $r \in L_{\infty e}$ and $\tau \in \mathbb{R}_+$, where Σr represents the system output.

Proof: Σ is finite-gain stable implies that $\forall x(0)$, $\exists c^* \geq 0$ and $\mu^* \geq 0$ such that

$$\|\Sigma r_\tau\| \leq \mu^* \|r_\tau\| + c^*$$

for all conceivable $r \in L_\infty$ and $\tau \in \mathbb{R}_+$. If $c \geq c^*/\mu^*$, then any $\mu \geq \mu^*$ will make (2-10) hold. For

the case $c < c^*/\mu^*$, it is true that $\forall c > 0$

$$\frac{\|\Sigma r_\tau\|}{\|r_\tau\| + c} \leq \frac{\mu^* \|r_\tau\| + c^*}{\|r_\tau\| + c} \leq \frac{c^*}{c}$$

The proof is complete.

$\square$

In section 2.3.1, we present the stability results for the general case in which all the sub-systems are nonlinear, and in section 2.3.2, the analysis is focused on the special case in which all the sub-systems are LTI. In the following analysis, we always assume that Σ_C is consistent with Σ_P and all the sub-systems are 0-detectable.

2.3.1 Nonlinear Sub-Systems

The stability results for the general case where all the sub-systems are nonlinear are summarized in the following theorem.

Theorem 2-1: Consider the control system in Figure 2-1, in which Σ_S is implemented with the hysteresis switching algorithm and the cost function defined in (2-8), and Σ_C contains at least one finite-gain stabilizing controller.

39

(i) The number of switches is bounded from the above.

(ii) If the matching controller state is assigned to each active controller when switched on, then the closed-loop system is stable.

(iii) More generally, if for each C_i ($i \in \mathbf{m}$), its matching reference signal generator is stable w.r.t. initial state, then the closed-loop system is stable.

(iv) Suppose for each candidate controllers $C_i = \{C_{1,i}, C_{2,i}\}$ ($i \in \mathbf{m}$) described by (2-7a,b), $C_{1,i}$ is output-input stable with $N=0$, $C_{2,i}$ is strongly stable, and the matching reference signal generator is stable w.r.t. initial state. If at each switching time t_k, the state for the active controller, $x_c(t_k)$, is set so that $\left| \tilde{x}_c(t_k) - x_c(t_k) \right|$ is bounded by some constant, where $\tilde{x}_c(t_k)$ is the matching controller state for the active controller at time t_k, then the closed-loop system is strongly stable. Furthermore, if for each $i \in \mathbf{m}$, $C_{2,i}$ is finite-gain stable and $C_{1,i}$ is output-input stable with a linear class $\mathcal{K}_\infty$ function in (2-3), then the closed-loop system is finite-gain stable.

(v) If the closed-loop system is stable, and all the sub-systems are 0-detectable, then all the states are bounded for $r \in L_\infty$.

Proof: (i) It follows Lemma 2-1 that the cost associated with the finite-gain stabilizing controller is finite, and then

$$\mu = \inf_{i \in \mathbf{m}} \left(\sup_{z \in \mathcal{Z}, r \in \mathbb{R}_+} \frac{\|z_\tau\|}{\left\| \left(\tilde{r}(C_i, z) \right)_\tau \right\| + c} \right)$$

is finite. Hence the number of switches is bounded from the above, which follows "Hysteresis Switching Lemma" in [54]. In this proof, we denote $\{t_k \mid k \in \mathbf{K}\}$ as an ordered sequence of switching times and $\hat{C}(t)$ as the active controller at time t.

(ii) Since there are a finite number of switches, we only need to consider the system stability after the final switch. Let us suppose the final switch happens at time t_n. it is true that

$$\sup_{\tau \in \mathbb{R}_+} \frac{\|z_\tau\|}{\left\|\left(\tilde{r}\left(\hat{C}(t_n),z\right)\right)_\tau\right\| + c} < \mu + \varepsilon \tag{2-11}$$

Let

$$c_0 = \left\|\left(\tilde{r}\left(\hat{C}(t_n),z\right)\right)_{[0,t_n)}\right\| \tag{2-12}$$

If the matching controller state is assigned to each active controller when switched on, then we have

$$r_{[t_n,\tau)} = \left(\tilde{r}\left(\hat{C}(t_n),z\right)\right)_{[t_n,\tau)}$$

for all $\tau \geq t_n$. Hence it is derived that

$$\|z_\tau\| < (\mu+\varepsilon)\|r_\tau\| + (\mu+\varepsilon)(c_0 + c)$$

for all $\tau \geq t_n$, which means the closed-loop system is stable.

(iii) If for each C_i ($i \in \mathbf{m}$), its matching reference signal generator is stable w.r.t. initial state, then there exists a positive number B such that

$$\left\|r_{[t_n,\tau)} - \left(\tilde{r}\left(\hat{C}(t_n),z\right)\right)_{[t_n,\tau)}\right\| \leq B$$

for all $\tau \geq t_n$, where t_n is the time of the final switch. Hence (2-11) implies that

$$\|z_\tau\| < (\mu + \varepsilon)\|r_\tau\| + (\mu + \varepsilon)(B + c_0 + c)$$

for all $\tau \geq t_n$, where c_0 is defined in (2-12), i.e. the closed-loop system is stable.

(iv) Since $C_{1,i}$ is output-input stable with $N=0$ for any $i \in \mathbf{m}$, there exist functions $\gamma_1 \in \mathcal{K}_\infty$

and $\beta \in \mathcal{KL}$ such that

$$\left| \tilde{e}_i (t) \right| \leq \gamma_1 \left(\left\| (u_c)_t \right\| \right) + \beta \left(\left| x_{c,i} (0) \right|, t \right) \Rightarrow \left\| (\tilde{e}_i)_t \right\| \leq \gamma_1 \left(\left\| (u_c)_t \right\| \right) + \beta \left(\left| x_{c1,i} (0) \right|, 0 \right)$$

for all $i \in \mathbf{m}$ and $t \in \mathbb{R}_+$. Since $C_{2,i}$ is strongly stable for any $i \in \mathbf{m}$, there exist a constant

$c_1 \geq 0$ and a function $\gamma_2 \in \mathcal{K}_\infty$ such that

$$\left\| (\tilde{v}_{c,i})_t \right\| \leq \gamma_2 \left(\left\| (y_s)_t \right\| \right) + c_1$$

for all $i \in \mathbf{m}$ and $t \in \mathbb{R}_+$. Hence we can find a constant $\bar{c} > 0$ and a function $\bar{\gamma} \in \mathcal{K}_\infty$ such

that

$$\left\| \left(\tilde{r}(C_i, z) \right)_\tau \right\| \leq \bar{\gamma} \left(\|z_\tau\| \right) + \bar{c} \tag{2-13}$$

holds for all $i \in \mathbf{m}$ and $t \in \mathbb{R}_+$. Since all the matching reference signal generators are stable

w.r.t. initial state, and at each switching time t_k, the state for the active controller, $x_c (t_k)$,

is set so that $\left| \tilde{x}_c (t_k) - x_c (t_k) \right|$ is bounded by some constant, where $\tilde{x}_c (t_k)$ is the matching controller state for the active controller at time t_k, there exists a positive number B such that

$$\left\| r_{[t_k, t_{k+1})} - \left(\tilde{r}\left(\hat{C}(t_k), z\right) \right)_{[t_k, t_{k+1})} \right\| \leq B \tag{2-14}$$

for all t_k ($k \in \mathbf{K}$) (If t_k is the time of the final switch, then $t_{k+1} = \infty$).

For a given set of reference input r and initial states, let $\{t_k \mid k \in \mathbf{K}\}$ be the switching times. It is true that

$$\sup_{\tau \in [0, t_1)} \frac{\|z_\tau\|}{\left\| \left(\tilde{r}\left(\hat{C}(0), z\right) \right)_\tau \right\| + c} < \mu + \varepsilon$$

for any $\tau \in [0, t_1)$. Hence we have

$$\|z_\tau\| \leq (\mu + \varepsilon)\|r_\tau\| + (\mu + \varepsilon)(B + c) \tag{2-15}$$

for any $\tau \in [0, t_1)$. For the controller active on the time interval $[t_1, t_2)$, we have

$$\sup_{\tau \in [0, t_2)} \frac{\|z_\tau\|}{\left\| \left(\tilde{r}\left(\hat{C}(t_1), z\right) \right)_\tau \right\| + c} < \mu + \varepsilon \tag{2-16}$$

Based on (2-13) and (2-15), we know that

$$\left\| \left(\tilde{r}\left(\hat{C}(t_1), z\right) \right)_{[0, t_1)} \right\| \leq \overline{\gamma}\left((\mu + \varepsilon)\|r_{t_1}\| + (\mu + \varepsilon)(B + c) \right) + \overline{c}$$

Together with (2-14), (2-16) leads to

$$
\begin{aligned}
\|z_\tau\| &< (\mu + \varepsilon)\left(\left\| \left(\tilde{r}\left(\hat{C}(t_1), z\right) \right)_{[0, t_1)} \right\| + \left\| \left(\tilde{r}\left(\hat{C}(t_1), z\right) \right)_{[t_1, \tau)} \right\| \right) + (\mu + \varepsilon)c \\
&\leq (\mu + \varepsilon)\left[\overline{\gamma}\left((\mu + \varepsilon)\|r_{t_1}\| + (\mu + \varepsilon)(B + c) \right) + \overline{c} + \left\| r_{[t_1, \tau)} \right\| + B \right] + (\mu + \varepsilon)c \\
&\leq (\mu + \varepsilon)\left[\overline{\gamma}\left((\mu + \varepsilon)\|r_{t_1}\| + (\mu + \varepsilon)(B + c) \right) + \left\| r_{[t_1, \tau)} \right\| \right] + (\mu + \varepsilon)(B + c + \overline{c})
\end{aligned}
$$

for any $\tau \in [t_1, t_2)$. Hence we can find a constant $\bar{c} > 0$ and a function $\bar{\gamma} \in \mathcal{K}_\infty$ so that

$$\|z_\tau\| \le \bar{\gamma} \left(\left\| r_{_{[0,\tau)}} \right\| \right) + \bar{c}$$

for any $\tau \in [t_1, t_2)$. Here we have abused the notations $\bar{c}$ and $\bar{\gamma}$. By repeating the above

procedure, we can always show that there exist a constant $\bar{c} > 0$ and a function $\bar{\gamma} \in \mathcal{K}_\infty$ so

that $\|z_\tau\| \le \bar{\gamma} \left(\left\| r_{_{[0,\tau)}} \right\| \right) + \bar{c}$ for any $\tau \in \mathbb{R}_+$. As we can see, the choices for $\bar{c}$ and $\bar{\gamma}$ at each

step are independent of the reference signal. Hence the closed-loop system is strongly

stable. Furthermore, if for each $i \in \mathbf{m}$, $C_{2,i}$ is finite-gain stable and $C_{1,i}$ is output-input

stable with a linear class $\mathcal{K}_\infty$ function in (2-3), then we can repeat the above proof

procedure and show that $\|z_\tau\| \le \bar{\gamma} \left\| r_{_{[0,\tau)}} \right\| + \bar{c}$ holds for all $\tau \in \mathbb{R}_+$ and $\bar{c}$, $\bar{\gamma}$ are positive

some numbers independent of the reference signal. Hence the closed-loop system is

shown to be finite-gain stable.

(v) It is straightforward to see that if the closed-loop system is stable and $r \in L_\infty$, then

$z \in L_\infty$, and then all the states are bounded since all the sub-systems are 0-dectectable.

$\square$

The results in parts (i) and (iii) of Theorem 2-1 have been shown with the L_{2e}

norm in [84, 85], while the other results in Theorem 2-1 have not appeared in any

literature. The result presented in (ii) is only theoretically existent when $C_{1,i}$ possesses its own state, since it is impossible to exactly set the matching controller state for $C_{1,i}$ in practice. It is interesting to notice that the stability results achieved in (iv) are stronger than that in (iii), i.e. the stability result presented in [84, 85], with the introduction of "matching controller state". No such strong stability results have been developed since resetting controller states is either not considered [84, 85] or performed only to ensure "bumpless" switch in the previous research [31],

Remark 2-7: In the cost function designed in (2-8), the $L_{\infty e}$ norm is employed since we are interested in the stability defined in the $L_{\infty e}$ norm. For the stability defined in any other norm, once the cost function is properly defined, similar stability results can be achieved with the same approach.

Example 2-1: Consider the plant Σ_P, which can be described as

$$\begin{cases} \dot{x}_p = f(x_p)x_p + h \cdot g(x_p) \cdot u_p \\ y_p = x_p \end{cases}$$

where h is unknown and takes value at either 1 or -1, $f(x_p)$ and $g(x_p)$ are two smooth functions and there exist two positive constants f_0 and g_0 such that $0 < g_0 \le g(x_p)$ and $|f(x_p)| \le f_0$. The disturbance terms are zeros. Suppose h is known, and then the following control law

$$u_p = h \cdot \left[\left(-1 - f(y_p) \right) \cdot y_p + r \right] / g(y_p)$$

is finite gain stabilizing. Since h is unknown, we can design two candidate controllers, one for $h=1$ and one for $h-1$. The matching signal generator for each controller is

$$r = h \cdot g\left(y_p\right) \cdot u_p + y_p + f\left(y_p\right) \cdot y_p$$

With the proposed switching algorithm, the resulted switching control system is finite-gain stable.

2.3.2 LTI Sub-Systems

In this sub-section, all the controllers in Σ_C are stabilizable and detectable LTI systems. For each $C_i=\{C_{1,i}, C_{2,i}\}$, $i\in\mathbf{m}$, $C_{1,i}$ and $C_{2,i}$ are both represented by proper transfer functions. Σ_P is a controllable and detectable LTI system whose input-output behavior can be described by a strictly proper transfer function. In this case, a stabilizing controller $C\in\Sigma_C$ guarantees that the closed-loop system in Figure 2-3 is finite-gain stable. For a general LTI controller of two degree-of-freedom shown in Figure 2-2, the matching reference signal generator shown in Figure 2-4 is unstable (w.r.t. initial state) when C_1 is non-minimum phase or C_2 is unstable. The adaptive scheme presented in [84, 85] may fail, since the adaptive control problem formulated in section 2.3 is not "feasible" (defined in [84, 85]). Furthermore, if C_1 is strictly proper, then the generated matching reference signals may contain impulses, which are not desired in numerical calculation. Hence we have to find a different way to construct the matching reference signal generator and the related cost function so that the non-identifier based adaptive strategy can work with general LTI controllers.

For any LTI controller of two degree-of-freedom $C=\{C_1,C_2\}$ shown in Figure 2-2, we can always construct a stabilizable and detectable LTI controller of three degree-of-freedom $\overline{C}=\{\overline{C}_1,\overline{C}_2,\overline{C}_3\}$, shown in Figure 2-6, that has the same input-output behavior as C, i.e. $\overline{C}_1\overline{C}_3=C_1$ and $\overline{C}_1\overline{C}_2=C_1C_2$. Furthermore, such a three degree-of-freedom controller can be constructed so that $\overline{C}_1$ is bi-proper and minimum phase, and $\overline{C}_2$ and $\overline{C}_3$ are proper and stable. For example, given a controller of two degree-of-freedom

$$C=\left\{\frac{s-1}{(s+1)(s+2)},\frac{10}{s+10}\right\}$$

the controller of three degree-of-freedom

$$\overline{C}=\left\{1,\frac{10(s-1)}{(s+10)(s+1)(s+2)},\frac{s-1}{(s+1)(s+2)}\right\}$$

possesses the same input-output characteristics as C does. It can be proven that when the unknown plant is LTI, $C=\{C_1,C_2\}$ is a stabilizing controller if and only if $\{\overline{C}_1,\overline{C}_2\}$ is stabilizing. For a controller $C=\{C_1,C_2\}$ in which C_1 is bi-proper and minimum phase and C_2 is proper and stable, we always set $\overline{C}_1=C_1$, $\overline{C}_2=C_2$ and $\overline{C}_3=1$. Now we propose a generalized matching reference signal generator for controller $C=\{C_1, C_2\}$ as shown in Figure 2-7. The output of this system, $\tilde{r}_g$, is the matching signal corresponding to r_g shown in Figure 2-6.

The cost function for a candidate controller $C\in\Sigma_C$ is then chosen as

$$V(C,z,t)=\sup_{\tau\in[0,t]}\frac{\|z_\tau\|}{\left\|\left(\tilde{r}_g(C,z)\right)_\tau\right\|+c} \tag{2-17}$$

It is trivial to see that if a candidate controller $C=\{C_1, C_2\}$ is stabilizing, then its cost

calculated by (2-17) is always bounded. For the trivial case where $\overline{C}_1 = C_1$, $\overline{C}_2 = C_2$ and

$\overline{C}_3 = 1$, the cost function in (2-17) is the same as in (2-8).

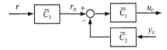

Figure 2-6: Diagram of a three degree-of-freedom controller.

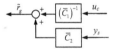

Figure 2-7: Generalized matching reference signal generator.

Before we present the stability results, let us recall the following lemma [39],

which states that a LTI system detectable in the classical sense is 0-detectable, i.e.

bounded input and output indicate bounded plant state.

Lemma 3-2 [39]: A LTI system is detectable in the classical sense indicates it is 0-

detectable.

Proof: Consider a general LTI system

$$\begin{cases} \dot{x} = Ax + Bu \\ y = Cx + Du \end{cases} \tag{2-18}$$

where A, B, C and D are constant matrices with proper dimensions. If it is detectable in the classical sense, then there exists a matrix L, such that $A+LC$ is Hurwitz. In this case, the system

$$\dot{x}_d = Ax_d + Bu + L\left(Cx_d + Du - y\right) \tag{2-19}$$

referred to as an observer, generates the observed state $x_d(t)$ such that the observation error $|x(t) - x_d(t)|$ converges to zero exponentially fast. In particular, if $x(0) = x_d(0)$, then $x(t) = x_d(t)$ for all non-negative t. Therefore

$$x\left(t, x_0, u\right) = e^{t(A+LC)} x_0 + \int_0^t e^{\tau(A+LC)} \left[\left(B + LD\right) u\left(t - \tau\right) - Ly\left(t - \tau\right) \right] d\tau$$

where $x_0(t) = x(0)$. Choose two positive numbers δ' and δ so that $\mathbf{Re}(\lambda) \le -\delta' < -\delta$ for every eigenvalue λ of $A+LC$. Then there exist a polynomial $P(\cdot)$ and a positive constant K such that

$$\left| x\left(t, x_0, u\right) \right| \le P(t) e^{-\delta t} \left| x_0 \right| + \int_0^t P(\tau) e^{-\delta \tau} \left[\left| u\left(t - \tau\right) \right| + \left| y\left(t - \tau\right) \right| \right] d\tau$$
$$\le K e^{-\delta t} \left| x_0 \right| + K \left\| u_t \right\| + K \left\| y_t \right\|$$

The proof is complete.

$\square$

Lemma 2-3: Consider the control system in Figure 2-1, in which d_u and d_y are bounded and piecewise continuous, Σ_P is LTI and detectable, Σ_C is a finite set of stabilizable and detectable LTI controllers, and Σ_S is implemented with the hysteresis switching algorithm

49

and the cost function is chosen as (2-13). If there exists one stabilizing controller in Σ_C and then the overall system is stable, and all the states are bounded when $r \in L_\infty$.

Proof: Since the cost associated with the stabilizing controller is finite, it is true that

$$\mu = \inf_{i \in \mathbf{m}} \left(\sup_{z \in \mathcal{Z}, \tau \in \mathbb{R}_+} \frac{\|z_\tau\|}{\left\|\left(\tilde{r}_g\left(C_i, z\right)\right)_\tau\right\| + c} \right)$$

is finite, and we can show that the number of switches is bounded from the above. Let us suppose that the final switch happens at time t_n and $\hat{C} \in \Sigma_C$ is the final active controller. Hence it is true that

$$\sup_{\tau \in \mathbb{R}_+} \frac{\|z_\tau\|}{\left\|\left(\tilde{r}_g\left(\hat{C}, z\right)\right)_\tau\right\| + c} < \mu + \varepsilon$$

Let

$$c_0 = \left\|\left(\tilde{r}_g\left(\hat{C}, z\right)\right)_{[0, t_n]}\right\|$$

Hence it is true that

$$\|z_\tau\| < (\mu + \varepsilon)\left\|\left(\tilde{r}_g\left(\hat{C}, z\right)\right)_\tau\right\| + (\mu + \varepsilon)(c_0 + c) \tag{2-20}$$

for all $\tau \geq t_n$. For the final active controller $\hat{C}$, we can construct a three-degree of freedom controller, as shown in Figure 2-6, to produce the same control effort u_c. Since $\bar{C}_1$ is bi-proper and minimum phase and $\bar{C}_2$ is proper and stable, we can show that

$$\lim_{t \to \infty}\left\{\left\|r_g(t) - \tilde{r}_g\left(\hat{C}, z, t\right)\right\|\right\} = 0 \tag{2-21}$$

Since $\bar{C}_3$ is stable, we can also show that there exist $\mu_1 \geq 0$ and $c_1 \geq 0$, such that

$$\left\|\left(r_g\right)_t\right\| \leq \mu_1 \|r_t\| + c_1 \tag{2-22}$$

for any $t \geq t_n$. Hence it can be derived based on (2-20)-(2-22) that the closed-loop system is stable. Furthermore, with Lemma 2-2, we can show that all the states are bounded when $r \in L_\infty$.

□

In the previous analysis, the controller output u_c is assumed to be precisely known [84, 85]. Let us now consider the case in which u_c is corrupted by a piecewise continuous and bounded disturbance term d in the path to Σ_S. In the case, (2-21) does not hold any more. However, the system stability can still be established by showing that there exists a function $\beta \in \mathcal{K}_\infty$, such that

$$\lim_{t \to \infty} \left\{ \left\| r_g(t) - \tilde{r}_g\left(\hat{C}, z, t\right) \right\| \right\} < \beta\left(\|d\|\right)$$

i.e., the difference between the real reference signal and the matching reference signal of the final active controller is bounded.

Now let us consider the simple case in which $C_{1,i}$ is bi-proper and minimum phase and $C_{2,i}$ is proper and stable so that the matching reference signal generator shown in Figure 2-4 is stable (w.r.t. initial state) and the achieved matching signals will not contain impulses. We show that the closed-loop system is guaranteed to be stable with the presence of bounded disturbances as long as there is a stabilizing candidate controller.

Furthermore, finite-gain stability can be achieved if the states for active controllers are properly set.

Lemma 2-4: Consider the control system in Figure 2-1, in which d_u and d_y are bounded and piecewise continuous, Σ_S is implemented with the hysteresis switching algorithm and the cost function defined in (2-8), Σ_C is a finite set of stabilizable and detectable LTI controllers and for any $C_i = \{C_{1,i}, C_{2,i}\}$ ($i \in \mathbf{m}$) $C_{1,i}$ is bi-proper and minimum phase and $C_{2,i}$ is proper and stable. If Σ_C contains one stabilizing controller, then the overall system is stable, and all the states are bounded when $r \in L_\infty$. Furthermore, if at each switching time t_k, the state for the active controller, $x_c(t_k)$, is set so that $\left| \tilde{x}_c(t_k) - x_c(t_k) \right|$ is bounded by some pre-specified constant, where $\tilde{x}_c(t_k)$ is the matching state for the active controller at time t_k, then the closed-loop system is finite-gain stable.

Proof: Since the all the sub-systems are LTI and d_u and d_y are bounded and piecewise continuous, it is easy to show that the stabilizing controller is finite-gain stabilizing. We can also show that the matching reference signal generator for each candidate controller is stable w.r.t. initial state. Hence Lemma 2-4 can be deduced from Lemma 2-1.

□

2.4 Concluding Remarks

The non-identifier based adaptive control system investigated in this chapter was originally formulated in [84, 85]. It can be viewed as a feedback interconnection of a SISO plant, a finite set of candidate controllers and a non-identifier based adaptive scheme, also referred to as supervisor or supervisory controller. The supervisory controller is designed without using any explicit assumptions on the plant model, and it guarantees system stability as long as certain plant independent requirements are satisfied and the candidate controller set contains one stabilizing controller. For the general case where all the sub-systems are nonlinear, we have shown that stronger stability properties can to be achieved if the state for each active controller can be reset properly. In addition, we have proposed a generalized matching reference signal generator so that the adaptive scheme can be applied with any general LTI controllers when the unknown plant is LTI. It is easy to check that similar results can be developed with other hysteresis-based switching algorithms discussed in [22].

In the analysis, the stability of the closed-loop system is established with the assumption that the candidate controller set contains at least one finite-gain stabilizing controller. Let us consider the situation in which the plant can be described by a family of admissible models and each of them can be stabilized by one and only one candidate controller. If the plant changes from one admissible model to another, the adaptive scheme discussed in this chapter may not guarantee system stability since no candidate controller is always stabilizing. To handle such a situation, it is intuitive to redesign the

53

supervisory controller with a cost function fading out old measurements. It is still under investigation how to establish system stability in such a case.

CHAPTER 3 : SAFE ADAPTIVE CONTROL FOR

PERFORMANCE IMPROVEMENT

3.1 Introduction

A generalized adaptive controller can be viewed as a parameterized controller whose control parameters are adjusted by certain adaptive strategy [51]. It can be classified as *identifier based* or *non-identifier based*, depending on whether or not certain model identifier is employed. In the past fifty years, many identifier based adaptive schemes have been developed with well established stability properties and extensive successful applications have been reported. However, applications of adaptive control in safety sensitive systems, such as high-performance aircraft, are still very limited. The major reason is that the identifier-based adaptive controllers suffer from the conflict between parameter estimation and control, and they may lead to worse transient performance than a non-adaptive controller when poor initial estimates are used. Large transient oscillations may be created by the adaptive controller to improve the estimation quality. Furthermore, when the aircraft flies from one point to another, its dynamics change drastically and the traditional identifier-based adaptive controller may not be able to tune the control parameters fast enough to guarantee good performance. Another drawback of the identifier-based adaptive control in addition to possible undesirable transients is that its nonlinear nature makes difficult to check stability bounds and predict transient performance as done in the linear time invariant (LTI) case. As a result in the applications where safety is at stake practitioners are reluctant to close the loop with an identifier-based adaptive controller. On the other hand, the popularly used control laws in safety sensitive systems, such as robust controllers and gain-scheduling techniques, do not account for unpredictable changes in the plant and may lead to deterioration of performance when control environment changes or system failures appear.

Suppose we have two candidate controllers available. One is a non-adaptive controller, which has been shown to be stabilizing by theoretical analysis and extensive experiments. The other is an identifier based adaptive controller, which has the potential of improving system performance when the modeling error is sufficiently small. It may lead to poor transient performance or even threaten the system stability when the model error is large. The question arises: Can we design a *supervisory controller* (also referred to as *supervisor*) to guide the switching process among the two controllers as shown in Figure 3-1, so that the closed-loop system is guaranteed to be stable and the adaptive candidate controller is switched on only when it can improve system performance? The switching controller formed by the supervisor and the two candidate controllers falls into the concept of generalized adaptive controller, and it is referred to as the *safe adaptive controller* since the adaptive candidate controller is activated in a conservative but safe manner. The significance of such a safe adaptive controller is that it will push the existing adaptive schemes into the application fields where their stability and performance are doubted.

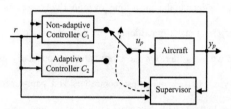

Figure 3-1: Switching between non-adaptive and adaptive candidate controllers.

In this chapter, we are going to address such a question in the model reference control (MRC) case. The two candidate controllers are designed using the classic methods presented in many textbooks such as [27], and they are briefly covered in this chapter. The supervisor involved in the safe adaptive control is a non-identifier based strategy inspired by the work presented in [65, 84, 85], which evaluates the performance of the two candidate controllers on-line without using any plant model. As we can see, the safe adaptive controller investigated in our work contains both identifier and non-identifier based adaptive schemes, and the control parameters are tuned in a piecewise continuous way. Hence when the plant changes slowly with time, the adaptive controller might be turned on to counteract the changes and provide good performance. When the adaptive controller deteriorates system performance, the robust non-adaptive controller will be turn on to guarantee system stability. It is shown in this chapter that the system stability is guaranteed when the supervisor is designed properly. Simulation results indicate that the proposed safe adaptive controller can achieve perform superior to that can be achieved by either of the candidate controllers. The rest of the chapter is organized as the follows. Problem formulation is given in section 3.2. In section 3.3, we introduce

57

the design of the supervisory controller and the stability results. For the completeness of the work, we also briefly review the control designs for MRC. In section 3.4, simulation results are presented to demonstrate the performance of the proposed safe adaptive control scheme. Concluding remarks are given in section 3.5.

3.2 Problem Formulation

3.2.1 Preliminaries

For a real function of time, $x(t)$, we define the L_∞ norm is defined as

$$\|x\|_\infty = \operatorname*{ess\,sup}_{\tau \in \mathbb{R}_+} \left(|x(\tau)| \right)$$

where $\mathbb{R}_+ = [0, \infty)$ and by *essential supremum* we mean

$$\operatorname*{ess\,sup}_{\tau \in \mathbb{R}_+} \left(|x(\tau)| \right) = \inf \left\{ a \,|\, |x(\tau)| \le a \text{ almost everywhere} \right\}$$

We say that $x \in L_\infty$ when $\|x\|_\infty$ is finite. We define the truncation of $x(t)$ over a time interval $I \subset \mathbb{R}_+$ as

$$x_I = \begin{cases} x(t), & \text{if } t \in I \\ 0, & \text{otherwise} \end{cases}$$

and in particular we use x_τ to denote the truncation of $x(t)$ over the time interval $[0, \tau)$.

For the functions of time that do not belong to L_∞, we define the $L_{\infty e}$ norm is defined as

$$\|x_t\|_\infty \triangleq \operatorname*{ess\,sup}_{0 \le \tau \le t} \left(|x(\tau)| \right)$$

Since in this chapter the stability is most interested in the L_∞ norm, $\|\cdot\|$ is used to represent $\|\cdot\|_\infty$ unless otherwise clarified.

In this chapter, we also consider the *exponentially weighted L_2 norm*, which is referred to as the $L_{2\delta}$ *norm*. The $L_{2\delta}$ norm is defined as

$$\|x_t\|_{2\delta} \triangleq \left(\int_0^t e^{-\delta(t-\tau)} x^T(\tau) x(\tau) d\tau \right)^{\frac{1}{2}}$$

where $\delta \geq 0$ is a constant. We say that $x \in L_{2\delta}$ when $\|x_t\|_{2\delta}$ exists for any finite t.

A continuous function $\alpha : \mathbb{R}_+ \rightarrow \mathbb{R}_+$ is said to be of class $\mathcal{K}$ if it strictly increasing and $\alpha(0) = 0$. If $\alpha \in \mathcal{K}$ is unbounded, then it is said to be of class $\mathcal{K}_\infty$. A function $\beta(s,t) : \mathbb{R}_+ \times \mathbb{R}_+ \rightarrow \mathbb{R}_+$ is said to be of class $\mathcal{KL}$, if $\beta(\cdot,t)$ is of class $\mathcal{K}$ for each fixed $t \geq 0$ and $\beta(s,t)$ decreases to 0 as $t \rightarrow \infty$ for each fixed $s \geq 0$.

To introduce the definition for stability, we consider a general finite-dimensional MIMO nonlinear system

$$\begin{cases} \dot{x} = f(x,u) \\ y = h(x,u) \end{cases} \tag{3-1}$$

where $x \in \mathbb{R}^n$, $u \in \mathbb{R}^m$ and $y \in \mathbb{R}^l$ are the system state, input and output (n, m and l are some positive integers), respectively, $f: \mathbb{R}^n \times \mathbb{R}^m \rightarrow \mathbb{R}^n$ and $h: \mathbb{R}^n \times \mathbb{R}^m \rightarrow \mathbb{R}^l$ are two smooth enough functions. Here, u is piecewise smooth and it represents only the input that we are interested in, i.e. the input that can be measured or controlled. Other inputs, such as the immeasurable disturbances, are implicitly included in (3-1).

Definition 3-1 (stability)

The system in (3-1) is said to be *(weakly) stable* if for any given initial condition $x(0) \in L_\infty$ and input $u \in L_{\infty e}$, there exist a non-negative constant c and a function $\gamma \in \mathcal{K}_\infty$ such that

$$\|y_t\| \leq \gamma(\|u_t\|) + c \tag{3-2}$$

holds for all $t \in \mathbb{R}_+$. Otherwise, it is said to be *unstable*. The system in (3-1) is said to be *strongly stable* if for any given initial condition $x(0) \in L_\infty$ there exist a non-negative constant c and a class $\mathcal{K}$ function γ such that (3-2) holds for all $u \in L_{\infty e}$ and $t \in \mathbb{R}_+$. In particular, if the system in (3-1) is strongly stable and the function γ in (3-2) is a linear function, then it is said to be *finite-gain stable*.

3.2.2 Problem Description

The safe adaptive control system investigated in this chapter can be viewed as a feedback interconnection of a SISO plant Σ_P, a candidate controller set Σ_C and a supervisory controller Σ_S (non-identifier based adaptive strategy), as shown in Figure 3-2, in which r is the reference signal, u_c is the controller output, u_p and y_p are the input and output of Σ_P, respectively, d_u and d_y are bounded piecewise continuous disturbance signals, y_s is the disturbed plant output, and σ is the piecewise constant switching signal generated by Σ_S.

60

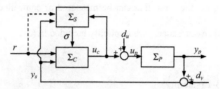

Figure 3-2: Diagram of the proposed safe adaptive control system.

We consider the case in which Σ_P is a linear SISO detectable system whose input-output characteristics can be presented as

$$y_p = G_p(s)u_p = G_0(s)\left[1+\Delta_m(s)\right]u_p \qquad (3\text{-}3)$$

where $G_p(s)$ is the transfer function of the plant, $G_0(s)$ is the nominal plant model, and $\Delta_m(s)$ is the modeling uncertainty. The candidate controller set Σ_C contains two candidate controllers C_i, $i \in \mathbf{m} = \{1, 2\}$, among which C_1 is a robust LTI model reference controller, and C_2 is an indirect robust model reference adaptive controllers. In this chapter, the robust controller C_1 is also referred to as the *nominal controller*. It is known that the closed-loop poles of the feedback interconnection formed by C_1 and Σ_P stay in $\mathbf{Re}(s) < -\lambda$ for some known $\lambda > 0$. We say C_1 is a stabilizing controller with *stability margin* λ. The design of the two candidate controllers will be presented in section 3.3.1. In the MRC case, the system performance is evaluated with the tracking error signal that is the difference between the plant output and the reference model output.

The supervisory controller to be design in this work, Σ_S, generates the switching signal σ with r, u_c, y_s. $\sigma : [0,\infty) \to \mathbf{m}$ is a piecewise constant function of time, and the

61

times at which the switching signal σ is discontinuous are called the switching times. We denote $\{t_k \mid k \in \mathbf{K}\}$ as an ordered sequence of switching times. For simplicity of analysis, we assume that σ is right continuous, i.e. if t_1 and t_2 are two consecutive switching times of a feasible signal σ, then σ is constant on $[t_1, t_2)$.

Though Σ_P is SISO, the closed-loop system in Figure 3-2 is considered as a single-input multi-output (SIMO) system, whose input is r and output vector is $[y_s, u_c]^T$. Here it is assumed that u_c and y_s can be perfectly measured, and we use z to represent the measured output vector $[y_s, u_c]^T$. We denote $\mathcal{Z}$ as the set of all conceivable signals of z.

The problem investigated in this chapter is described as the follow. For the control system given in Figure 3-2, design the supervisor Σ_S so that the resulted switching controller can lead to better performance than either of the candidate controllers when used alone.

Remark 3-1: The problem of using multiple models and switching to improve transient response investigated by Narendra and Balakrishnan in [57] looks similar to but differs from our problem. In their problem, it is known that a single adaptive controller along can assure stability though multiple adaptive controllers are employed. In our problem, we know that the non-adaptive controller guarantees system stability and then we design the supervisory controller to switch on the adaptive controller only if it can improve system performance. The supervisory controller is designed without the plant model. The system stability is guaranteed even if the candidate adaptive controller is destabilizing.

Remark 3-2: In [74], "safe adaptive control" refers to the non-identifier based adaptive investigated in Chapter 2. It is different from the "safe adaptive control" problem investigated in Chapter 3. In our problem, we have the pre-knowledge that the non-adaptive candidate controller is always stabilizing, and the adaptive candidate controller is switched on in a safe but conservative way to improve system performance.

3.3 MRC Designs

Let us consider the linear SISO detectable system in (3-3). The disturbed plant can be described by

$$y_s = G_0(s)\left[1 + \Delta_m(s)\right](u_c + d_u) + d_y \tag{3-4}$$

where $G_0(s)$ and $G_0(s)\Delta_m(s)$ are strictly proper. We design MRC and MRAC schemes with the nominal plant model of the form

$$y_s = G_0(s)u_c = k_p \frac{Z_p(s)}{R_p(s)} u_c \tag{3-5}$$

where $Z_p(s)$ and $R_p(s)$ are monic polynomials and k_p is a constant referred to as the high frequency gain. These schemes, however, have to be robust so that the closed-loop system is stable with the presence of modeling uncertainties and disturbance [27]. The transfer function of the reference model given by

$$y_m = W_m(s)r = k_m \frac{Z_m(s)}{R_m(s)} r \tag{3-6}$$

where $Z_m(s)$ and $R_m(s)$ are monic polynomials and k_m is a constant. The MRC objective is to determine the plant input so that all signals are bounded and the plant output tracks the reference model output y_m as close as possible.

In order to meet the MRC objective with a control law that is proper and use only measurable signals, the following assumptions have to hold [27].

Plant Assumptions:

P1. $Z_p(s)$ is a monic Hurwitz polynomial.

P2. An upper bound n of the degree n_p of $R_p(s)$ is known.

P3. The relative degree $n^* = n_p - m_p$ of $G_0(s)$ is known, where m_p is the degree of $Z_p(s)$.

P4. The sign of the high frequency gain k_p is known.

Reference Model Assumptions:

M1. $Z_m(s), R_m(s)$ are monic Hurwitz polynomials of degree q_m, p_m, respectively, where $p_m \leq n$.

M2. The relative degree of $W_m(s)$ is the same as that of $G_0(s)$, i.e., $n^* = p_m - q_m$.

3.3.1 Non-Adaptive MRC

For the simple case in which all the coefficients in $G_0(s)$ are known, let us consider the feedback control law [27]

$$u_c = \theta_1^T \omega_1 + \theta_2^T \omega_1 + \theta_3 y_s + c_0 r \qquad (3\text{-}7)$$

shown in Figure 3-3, where

$$\omega_1 = \frac{\alpha(s)}{\Lambda(s)} u_c, \quad \omega_2 = \frac{\alpha(s)}{\Lambda(s)} y_s$$

$$\alpha(s) \triangleq \alpha_{n-2}(s) = \left[s^{n-2}, s^{n-3}, \cdots, s, 1 \right]^T \quad \text{for } n \geq 2$$

$$\alpha(s) \triangleq 0 \qquad\qquad\qquad\qquad \text{for } n = 1$$

$c_0, \theta_3 \in \mathbb{R}^1$, $\theta_1, \theta_2 \in \mathbb{R}^{n-1}$ are constant parameters to be designed and $\Lambda(s)$ is an arbitrary

monic Hurwitz polynomial of degree $n-1$ that contains $Z_m(s)$ as a factor, i.e.,

$$\Lambda(s) = \Lambda_0(s) Z_m(s)$$

which implies that $\Lambda_0(s)$ is monic, Hurwitz and of degree $n_0 = n - 1 - q_m$.

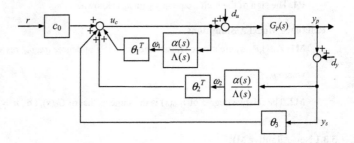

Figure 3-3: Structure of the MRC scheme.

The controller parameter vector

$$\Theta = \left[\theta_1^T, \theta_2^T, \theta_3, c_0 \right]^T \in \mathbb{R}^{2n}$$

is to be chosen so that the transfer function from r to y_s (in the nominal model) is equal to $W_m(s)$. We can choose

$$c_0 = k_m / k_p \tag{3-8a}$$

and find the solution for θ_i^*, $i = 1,2,3$ by inspecting

$$\theta_1^T \alpha(s) = \Lambda(s) - Z_p(s) Q(s) \tag{3-8b}$$

$$\theta_2^T \alpha(s) + \theta_3 \Lambda(s) = \frac{Q(s) R_p(s) - \Lambda_0(s) R_m(s)}{k_p} \tag{3-8c}$$

where $Q(s)$ (of degree $n-1-q_m$) is the quotient of $Z_p \Lambda_0 R_m / R_p$.

3.3.2 Robust MRAC

In the case where the coefficients in the nominal model $G_0(s)$ are unknown, we can use robust adaptive laws to estimate the coefficients and calculate the control parameters on-line. Here we briefly introduce the robust adaptive law with dynamic normalization and leakage modification based on the gradient method, and detailed information for robust adaptive laws can be found in [27]. To implement the robust adaptive law for parameter estimation, we need the additional assumption on the model uncertainty $\Delta_m(s)$:

A1. $\Delta_m(s)$ is analytical in $\text{Re}[s] \geq -\delta_0/2$ for some known $\delta_0 > 0$.

Let us assume that the nominal model $G_0(s)$ can be expressed as

$$G_0(s) = \frac{Z_p(s)}{R_p(s)} = \frac{b_m s^m + b_{m-1} s^{m-1} + \cdots + b_0}{s^n + a_{n-1} s^{n-1} + \cdots + a_0} u_p$$

where $b_m = k_p$ is the high frequency gain. For the disturbed plant described by (3-4), the robust adaptive law based on the gradient method is summarized in Table 3-1.

Plant	$y_s = G_0(s)\left[1 + \Delta_m(s)\right](u_c + d_u) + d_y$
Parametric Model	$z = \theta_p^{*T}\phi + \eta$ $$\theta_p^* = \left[b_m, \cdots, b_0, \ a_{n-1}, \cdots a_0\right]^T, \ \phi = \left[\frac{\alpha_m^T(s)}{\Lambda_p(s)}u_c, \ -\frac{\alpha_{n-1}^T(s)}{\Lambda_p(s)}y_s\right]^T,$$ $$z = \frac{s^n}{\Lambda_p(s)}y_s, \ \eta = \frac{Z_p(s)}{\Lambda_p(s)}\Delta_m u_c + \frac{Z_p(s)}{\Lambda_p(s)}\left(1 + \Delta_m(s)\right)d_u + \frac{R_p(s)}{\Lambda_p(s)}d_y$$
Adaptive Law	$\dot{\theta}_p = \text{Proj}\left\{\Gamma\left[\varepsilon\phi - \sigma(\theta - \theta_0)\right]\right\}, \ \varepsilon = \dfrac{z - \theta_p^T\phi}{m^2}$
Normalization Term	$m^2 = 1 + \phi^T\phi + n_s, \ \dot{n}_s = -\delta_0 n_s + u_c^2 + y_s^2, \ n_s(0) = 0$
Design Variables	$\Lambda_p(s) = s^n + \lambda_p^T\alpha_{n-1}(s)$: Monic Hurwitz of degree n; Proj$\{\cdot\}$ is the projection function to limit θ_p into a bounded convex set. $\sigma > 0$; $\Gamma = \Gamma^T > 0$; θ_0 is the best initial guess for θ

Table 3-1: Robust adaptive law based on gradient method.

The indirect robust MRAC can be implemented as the control law in (3-7) and the control parameters $\theta_1, \theta_2, \ \theta_3$ and c_0 can be solved by investigating the equations (3-8a)-(3-8c) at each time t with the on-line estimates achieved by the robust adaptive law in Table 3-1. The projection function used in the adaptive law prevents the estimate of b_m from crossing zero and causing a problem in calculating c_0 on-line. If the assumptions **P1-P4** and **A1** are satisfied, then the robust adaptive guarantees the estimates for unknown parameters are bounded and the closed-loop system is stable if the modeling uncertainty is sufficiently small [27]. If one of the assumptions fails, then the indirect adaptive controller may not be stabilizing. However, θ_p is always bounded due to the use of the projection function, which indicates that the calculated vector of control parameters

$\Theta = \left[\theta_1^T, \theta_2^T, \theta_3, c_0 \right]^T$ is always bounded. This property is important for the stability analysis in the next section.

In this section we only present an indirect MRAC scheme. Other MRAC schemes, no matter direct or indirect, can also be implemented in the safe adaptive control algorithm presented in the next section as long as that the calculated control gains are bounded and c_0 is prevented from crossing zero.

3.4 Design of the Supervisory Controller

In this section, we want to investigate how the supervisory controller Σ_S can be designed to improve system performance in the MRC case. The controller set Σ_C contains two candidate controllers. The nominal controller, C_1, is a LTI robust stabilizing controller, while the other candidate controller, C_2, is an indirect robust model reference adaptive controller. These two controllers can both be presented in the form in (3-7), which means they can be built to share the states ω_1 and ω_2 (defined in (3-7)). When the supervisory controller switches on C_i ($i \in \mathbf{m}$), it only needs to switch the corresponding control parameters $\Theta_i = \left[\theta_{1,i}^T, \theta_{2,i}^T, \theta_{3,i}, c_{0,i} \right]^T$ into the control system shown in Figure 3-3. It is not necessary to reset the controller state at each switching time. Hence the control effect u_c is generated as

$$u_c = \theta_{1,\sigma}^T \frac{\alpha(s)}{\Lambda(s)} u_c + \theta_{2,\sigma}^T \frac{\alpha(s)}{\Lambda(s)} y_s + \theta_{3,\sigma} y_s + c_{0,\sigma} r$$

as shown in Figure 3-3, where σ is the switching signal generated by the supervisory controller.

The supervisory controller designed in this section is motivated by the research in [65, 85], and it is a non-identifier based scheme, which evaluates the performance of candidate controllers without using any plant model. The supervisory controller is composed of three subsystems:

1. *Matching reference signal generator*: A dynamic system whose inputs are the on-line data u_c, y_s, ω_1 and ω_2, and whose outputs are the matching reference signals $\tilde{r}_i$, $i \in \mathbf{m}$.

2. *Controller cost generator*: a dynamic system whose inputs are the on-line data u_c and y_s and the matching reference signals $\tilde{r}_i$, $i \in \mathbf{m}$, and whose outputs V_i (cost for C_i), $i \in \mathbf{m}$ are suitable defined controller costs.

3. *Switching logic*: a dynamic system whose inputs are the on-line data u_c and y_s, the reference signal r and the control costs V_i, $i \in \mathbf{m}$ and whose output is a piecewise constant switching signal σ, taking values in $\mathbf{m}$, which is used to define the control law $u = u_\sigma$.

The matching reference signal generator for a candidate controller $C \in \Sigma_C$ is shown in Figure 3-4. For the adaptive controller, the control parameters θ_1, θ_2, θ_3 and c_0 are time varying. It is easy to find out that for either candidate controller, its matching reference signal is the same as the true reference signal on its active time intervals.

The cost function for a controller $C \in \Sigma_C$ is chosen as

69

$$V(C,z,t) = \frac{\left\| (y_s - \tilde{y}_m(C,z))_t \right\|_{2\delta} + w \left\| (u_c)_t \right\|_{2\delta}}{\left\| (\tilde{r}(C,z))_t \right\|_{2\delta} + c}$$ (3-9)

where w, c and δ are positive design constants, and $\tilde{y}_m = W_m \tilde{r}$. We may also use $\tilde{y}_{m,i}$ to

denote $\tilde{y}_m(C_i,z)$ when there is no confusion. We define the system cost as

$$V_s(z,t) = \frac{\left\| (y_s - y_m)_t \right\|_{2\delta} + w \left\| (u_c)_t \right\|_{2\delta}}{\left\| r_t \right\|_{2\delta} + c}$$ (3-10)

where w, c and δ are the same constants as in (3-9).

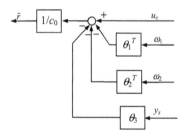

Figure 3-4: Matching reference signal generator for one candidate controller.

The switching logic, shown in Figure 3-5, is a modified hysteresis switching

algorithm with dwell time. Here τ is a timing signal and τ_D is a pre-specified positive

number called dwell time. By introducing the dwell time, we eliminate the possibility of

unbounded switching frequency. V_0 is a pre-specified positive number, which guarantees

that the nominal controller will be switched off only if the system cost is small enough.

70

$\rho : \mathbb{R}^n_+ \to \mathbf{m}$ denote the function whose value at $V = \{V_1, ..., V_m\}^T \in \mathbb{R}^m_+$, denoted as $\rho(V)$, is the least integer $i \in \mathbf{m}$ for which $V_i \le V_j$, $j \in \mathbf{m}$. $\varepsilon_1 \ge \varepsilon_2 > 0$ are two hysteresis constants in the switching logic.

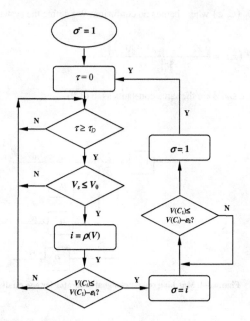

Figure 3-5: Modified hysteresis switching algorithm with dwell time.

Lemma 3-1: For a general LTI system

$$\begin{cases} \dot{x} = Ax + Bu \\ y = Cx \end{cases} \qquad (3\text{-}11)$$

71

where A, B, C and D are constant matrices with proper dimensions, given an arbitrary $\delta \geq 0$ there exist some constants $\gamma_1, \gamma_2 > 0$ and a function $\beta \in \mathcal{KL}$ such that for every initial state $x(0)$ and every u the corresponding solution satisfies the inequality

$$\left| y(t) \right| \leq \beta \left(\left| x(0) \right|, t \right) + \gamma_1 \left\| u_t \right\|_{2\delta} + \gamma_2 \left\| y_t \right\|_{2\delta}$$

as long as it exists.

Proof: It is well known that there exists a nonsingular matrix Q, such that the equivalence transformation $\overline{x} = Qx$ will transform (3-11) into

$$\begin{cases} \begin{bmatrix} \dot{\overline{x}}_o \\ \dot{\overline{x}}_{\overline{o}} \end{bmatrix} = \begin{bmatrix} \overline{A}_o & 0 \\ \overline{A}_{21} & \overline{A}_{\overline{o}} \end{bmatrix} \begin{bmatrix} \overline{x}_o \\ \overline{x}_{\overline{o}} \end{bmatrix} + \begin{bmatrix} \overline{B}_o \\ \overline{B}_{\overline{o}} \end{bmatrix} u \\ y = \begin{bmatrix} \overline{C}_o & 0 \end{bmatrix} \begin{bmatrix} \overline{x}_o \\ \overline{x}_{\overline{o}} \end{bmatrix} \end{cases}$$

where $\overline{x}_o$ and $\overline{x}_{\overline{o}}$ are the observable and unobservable states, respectively. For any $\delta > 0$, there exists a matrix L such that $\overline{A}_o + L\overline{C}_o$ is Hurwitz and $\mathbf{Re}(\lambda) < -\delta/2$ for every eigenvalue λ of $\overline{A}_o + L\overline{C}_o$. For the observable sub-system

$$\begin{cases} \dot{\overline{x}}_o = \overline{A}_o \overline{x}_o + \overline{B}_o u \\ y = \overline{C}_o \overline{x}_o \end{cases} \tag{3-12}$$

we can design the observer as

$$\dot{\tilde{x}}_o = \overline{A}_o \tilde{x}_o + \overline{B}_o u + L\left(\overline{C}_o \tilde{x}_o - y \right)$$

If $\tilde{x}_o(0) = \overline{x}_o(0)$, then $\tilde{x}_o(t) = \overline{x}_o(t)$ for all non-negative t. Therefore

$$\overline{x}_o\left(t, \overline{x}_o(0), u \right) = e^{t\left(\overline{A}_o + L\overline{C}_o \right)} \overline{x}_o(0) + \int_0^t e^{\tau\left(\overline{A}_o + L\overline{C}_o \right)} \left[\overline{B}_o u(t - \tau) - Ly(t - \tau) \right] d\tau$$

Let us choose a positive numbers δ' such that $\mathrm{Re}(\lambda) \leq -\delta' < -\delta$ for every eigenvalue λ of $\bar{A}_o + L\bar{C}_o$. Then there exist a polynomial $P(\cdot)$ such that

$$\left|\bar{x}_o\left(t, \bar{x}_o(0), u\right)\right| \leq P(t)e^{-\frac{\delta'}{2}t}\left|\bar{x}_o(0)\right| + \int_0^t P(\tau)e^{-\frac{\delta'}{2}\tau}\left[\left|u(t-\tau)\right| + \left|y(t-\tau)\right|\right]d\tau$$

$$\leq P(t)e^{-\frac{\delta'}{2}t}\left|\bar{x}_o(0)\right| + \int_0^t P(t-\tau)e^{-\frac{(\delta'-\delta)}{2}(t-\tau)}e^{-\frac{\delta}{2}(t-\tau)}\left|u(\tau)\right|d\tau$$

$$+ \int_0^t P(t-\tau)e^{-\frac{(\delta'-\delta)}{2}(t-\tau)}e^{-\frac{\delta}{2}(t-\tau)}\left|y(\tau)\right|d\tau$$

Applying the Schwartz inequality, we get

$$\int_0^t P(t-\tau)e^{-\frac{(\delta'-\delta)}{2}(t-\tau)}e^{-\frac{\delta}{2}(t-\tau)}\left|u(\tau)\right|d\tau \leq \left(\int_0^t P^2(t-\tau)e^{-(\delta'-\delta)(t-\tau)}d\tau\right)^{\frac{1}{2}}\left\|u_t\right\|_{2\delta}$$

$$\int_0^t P(t-\tau)e^{-\frac{(\delta'-\delta)}{2}(t-\tau)}e^{-\frac{\delta}{2}(t-\tau)}\left|y(\tau)\right|d\tau \leq \left(\int_0^t P^2(t-\tau)e^{-(\delta'-\delta)(t-\tau)}d\tau\right)^{\frac{1}{2}}\left\|y_t\right\|_{2\delta}$$

Hence there exists a positive constant K such that

$$\left|\bar{x}_o\left(t, \bar{x}_o(0), u\right)\right| \leq Ke^{-\frac{\delta'}{2}t}\left|\bar{x}_o(0)\right| + K\left\|u_t\right\|_{2\delta} + K\left\|y_t\right\|_{2\delta} \tag{3-13}$$

Together with $y = \bar{C}_o \bar{x}_o$ and $\bar{x} = Qx$, the equality in (3-13) indicates that there exist some constants $\gamma_1, \gamma_2 > 0$ and a function $\beta \in \mathcal{KL}$ such that the inequality

$$\left|y(t)\right| \leq \beta\left(\left|x(0)\right|, t\right) + \gamma_1\left\|u_t\right\|_{2\delta} + \gamma_2\left\|y_t\right\|_{2\delta}$$

always holds .

$\square$

Theorem 3-1: Consider the control system in Figure 3-1, in which the plant Σ_P is a detectable system described by (3-3), d_u and d_y are piecewise continuous and bounded, Σ_C contains one LTI stabilizing controller with a stability margin of $\delta/2$ and an indirect robust model reference controller whose control parameters are calculated based on the plant parameter estimates provided by the adaptive law in Table 3-1, and Σ_S is implemented with the modified hysteresis switching algorithm shown in Figure 3-5 and the controller cost function defined in (3-9). If the Hurwitz polynomial $\Lambda(s)$ used in (3-7) is designed such that all the zeros of $\Lambda(s) = 0$ stay in $\mathbf{Re}(s) < -\delta/2$ and the reference model W_m is chosen with all its poles in $\mathbf{Re}(s) < -\delta/2$, then the closed-loop system is finite-gain stable.

Proof: Let us denote the output vector of the closed-loop system as $z = [y_s, u_c]^T$. Since the adaptive law in Table 3-1 guarantees the estimates for the unknown plant parameters are bounded, it can be easily shown that the control parameters in C_2 are also bounded. Since $1/\Lambda(s)$ is analytical in $\mathbf{Re}(s) \geq -\delta/2$, there exists a constant $K_0 > 0$, such that

$$\left\| (\tilde{r}_i)_\tau \right\|_{2\delta} \leq K_0 \left\| z_\tau \right\|_{2\delta} + K_0 \tag{3-14}$$

holds for all $i \in \mathbf{m}$, $z \in \mathbf{Z}$ and $\tau \in \mathbb{R}_+$. Since C_1 is a stabilizing controller with a stability margin $\delta/2$ and W_m has all its poles in $\mathbf{Re}(s) < -\delta/2$, it is easy to show that the cost associated with C_1 is bounded and we set $\mu = \sup\limits_{z \in \mathbf{Z}, \tau \in \mathbb{R}_+} V(C_1, z, t)$. In the proof, we consider r is an arbitrary reference input and denote $\{t_k | k \in \mathbf{K}\}$ as an ordered sequence of switching times.

(i) For the time interval $[0, t_1)$, C_1 is active. C_1 is a stabilizing controller with a stability margin $\delta/2$ indicates that there exist some constants $\gamma_0 > 0$ and $c_0 > 0$, such that

$$\left\|z_\tau\right\|_{2\delta} < \gamma_2 \left\|r_\tau\right\|_{2\delta} + c_2 \tag{3-15}$$

for all $\tau \in [0, t_1)$.

(ii) Let us consider the system behavior when C_2 is active, i.e. during the time intervals $[t_k, t_{k+1})$ for $k=1,3,5,\dots$. It is true that

$$\frac{\left\|\left(y_s - W_m \tilde{r}_2\right)_\tau\right\|_{2\delta} + w \left\|\left(u_c\right)_\tau\right\|_{2\delta}}{\left\|\left(\tilde{r}_2\right)_\tau\right\|_{2\delta} + c} < \mu + \varepsilon_2 \tag{3-16}$$

for any $\tau \in [t_k, t_{k+1})$. From the switching logic, we also know that

$$\frac{\left\|\left(y_s - W_m r\right)_{t_k}\right\|_{2\delta} + w \left\|\left(u_c\right)_{t_k}\right\|_{2\delta}}{\left\|r_{t_k}\right\|_{2\delta} + c} \leq V_0$$

which indicates there exists a positive number k_2 such that

$$\left\|z_{t_k}\right\|_{2\delta} < k_2 \left\|r_{t_k}\right\|_{2\delta} + k_2 \tag{3-17}$$

for any $k=1,3,5,\dots$. In addition, we have that

$$\left(\tilde{r}_2\right)_{[t_k,\tau)} = r_{[t_k,\tau)} \tag{3-18}$$

for any $\tau \in [t_k, t_{k+1})$. With the knowledge that W_m is analytical in $\mathbf{Re}(s) \geq -\delta/2$, it can be derived from (3-14), (3-17) and (3-18) that there exists a positive number K_2 such that

$$\left\|\left(\tilde{r}_2\right)_\tau\right\|_{2\delta} \leq K_2 \left\|r_\tau\right\|_{2\delta} + K_2 \tag{3-19}$$

for any $\tau \in [t_k, t_{k+1})$. With (3-16) and (3-19), we can show that there some constants $\gamma_2 > 0$ and $c_2 > 0$, such that

$$\left\|z_\tau\right\|_{2\delta} < \gamma_2 \left\|r_\tau\right\|_{2\delta} + c_2 \tag{3-20}$$

for any $\tau \in [t_k, t_{k+1})$.

(iii) Let us consider the system behavior on the time intervals $[t_k, t_{k+1})$ for $k=2,4,6,\ldots$. It is true that

$$\frac{\left\|\left(y_s - W_m \tilde{r}_1\right)_\tau\right\|_{2\delta} + w\left\|\left(u_c\right)_\tau\right\|_{2\delta}}{\left\|\left(\tilde{r}_1\right)_\tau\right\|_{2\delta} + c} \leq \mu \tag{3-21}$$

for all $\tau \in \mathbb{R}_+$. From (3-20) we know that

$$\left\|z_{t_k}\right\|_{2\delta} < \gamma_2 \left\|r_{t_k}\right\|_{2\delta} + c_2 \tag{3-22}$$

any $k=2,4,6,\ldots$. Also we have that

$$\left(\tilde{r}_1\right)_{[t_k,\tau)} = r_{[t_k,\tau)} \tag{3-23}$$

for any $\tau \in [t_k, t_{k+1})$. With the knowledge that W_m is analytical in $\mathbf{Re}(s) \geq -\delta/2$, it can be derived from (3-14), (3-22) and (3-23) that there exists a positive number K_1 such that

$$\left\|\left(\tilde{r}_1\right)_\tau\right\|_{2\delta} \leq K_1 \left\|r_\tau\right\|_{2\delta} + K_1 \tag{3-24}$$

for any $\tau \in [t_k, t_{k+1})$. With (3-21) and (3-24), we can show that there some constants $\gamma_1 > 0$ and $c_1 > 0$, such that

$$\left\|z_\tau\right\|_{2\delta} < \gamma_1 \left\|r_\tau\right\|_{2\delta} + c_1 \tag{3-25}$$

for any $\tau \in [t_k, t_{k+1})$.

With (3-15), (3-20) and (3-25), we have shown that

$$\left\|z_\tau\right\|_{2\delta} < \gamma \left\|r_\tau\right\|_{2\delta} + c \tag{3-26}$$

holds for all $\tau \in \mathbb{R}_+$, where $\gamma = \max\{\gamma_0, \gamma_1, \gamma_2\}$ and $c = \max\{c_0, c_1, c_2\}$.

(iv) It is easy to show that

$$\|r_\tau\|_{2\delta} \le \frac{1}{\delta}\|r_\tau\|$$

holds for all piecewise continuous signal $r \in L_{\infty e}$ and $\tau \in \mathbb{R}_+$. Hence (3-26) leads to

$$\|z_\tau\|_{2\delta} < \frac{\gamma}{\delta}\|r_\tau\| + c$$

for all $\tau \in \mathbb{R}_+$. With Lemma 3-1, we know that for a given initial plant state there exists a

positive number h_1 such that

$$\left|y_p(\tau)\right| \le h_1 + h_1\left\|(u_p)_\tau\right\|_{2\delta} + h_1\left\|(y_p)_\tau\right\|_{2\delta}$$

holds for all $\tau \in \mathbb{R}_+$. Since the disturbances are bounded, we have

$$\left|y_s(\tau)\right| \le h_1 + \|d_y\| + h_1\left(\left\|(u_c)_\tau\right\|_{2\delta} + \frac{1}{\delta}\|d_u\|\right) + h_1\left(\left\|(y_s)_\tau\right\|_{2\delta} + \frac{1}{\delta}\|d_y\|\right) \tag{3-27}$$

holds for all $\tau \in \mathbb{R}_+$. Since the control effort is computed as

$$u_c = \theta_{1,\sigma}^T \frac{\alpha(s)}{\Lambda(s)} u_c + \theta_{2,\sigma}^T \frac{\alpha(s)}{\Lambda(s)} y_s + \theta_{3,\sigma} y_s + c_{0,\sigma} r$$

in which $1/\Lambda(s)$ is analytical in $\mathbf{Re}(s) \ge -\delta/2$ and the control parameters are bounded, it

can be shown that there exists a positive constant h_2 such that

$$\left|u_c(\tau)\right| \le h_2\left(\left\|(u_c)_\tau\right\|_{2\delta} + \left\|(y_s)_\tau\right\|_{2\delta} + \left|y_s(\tau)\right|\right) + h_2 \tag{3-28}$$

holds for all $\tau \in \mathbb{R}_+$. Based on (3-27) and (3-28), we can show that there exists a positive

constant H such that

$$\left| u_c (\tau) \right| + \left| y_s (\tau) \right| \leq H \left\| r_\tau \right\| + H$$

holds for all $\tau \in \mathbb{R}_+$. Hence is true that

$$\left\| z_\tau \right\| \leq \left\| (u_c)_\tau \right\| + \left\| (y_s)_\tau \right\| \leq H \left\| r_\tau \right\| + H$$

holds for all $\tau \in \mathbb{R}_+$, i.e. the closed-loop system is finite gain stable.

□

Remark 3-3: In the stability proof, we do not use any assumption on the plant model and the adaptive controller is not assumed to be stabilizing. In other words, even if the adaptive candidate controller is designed based on some wrong assumptions, the system stability can always be achieved as long as the nominal controller is stabilizing.

The supervisory controller designed in this chapter can be easily extended to the case where multiple adaptive candidate controllers are employed. In this case, the estimators for the plant parameters can be started with different initial values to improve system performance as done in [57].

3.5 Simulation Results

In the simulations, the plant is a detectable system whose input-output behavior can be described by the model

$$y_p = \frac{b}{s-a}\left(1+\Delta_m(s)\right)\left(u_c + d_u\right)$$

and we know that the nominal values for the a and b are $a_0 = 1$ and $b_0 = 1$. The control objective is to design u_c so that y_p can track the output of the following reference model

$$y_m = \frac{2}{s+2}r$$

as close as possible. In the simulations, we will set d_y as zero.

Based on the nominal model, we design the nominal controller C_1 as

$$u_{c,1} = -3y_p + 2r$$

Suppose that we know the uncertain term $\Delta_m(s)$ is sufficiently small so that the nominal controller has a stability margin of 0.5. The adaptive candidate controller is designed as

$$u_{c,2} = -\frac{a_m + \hat{a}}{\hat{b}}y_p + \frac{b_m}{\hat{b}}r$$

where $\hat{a}$ and $\hat{b}$ are on-line estimates for a and b generated with the adaptive law in Table 3-1. We use the projection function so that $-10 \le \hat{a} \le 10$ $0.1 \le \hat{b} \le 10$. The parameters in the cost function are chosen as $w = 0.001$, $c = 0.001$, $\delta = 1$. The parameters in the switching logic are chosen as $\varepsilon_1 = \varepsilon_2 = 0.05$, $V_0 = 10$, $\tau_D = 2$ second.

Simulation 1

In this simulation, $\Delta_m(s) = 0$, $d_u = 0$, and the reference signal and the true plant parameters are shown in Figures 3-6(a) and 3-6(b), respectively. The parameters involved in the adaptive law are chosen as $\Lambda_p(s) = s+1$, $\sigma = 10^{-5}$, $\Gamma = \text{diag}\{10, 10\}$ and the normalization term is $m^2 = 1 + 0.1\phi^T\phi + 0.01n_s$. When we apply the two candidate controllers individually in the control system, their error signals ($e_1 = y_p - y_m$) are shown in Figure 3-6(c). Applying the safe adaptive controller developed in this chapter, we get the error signal shown in Figure 3-6(d) and the switching signal is shown in Figure 3-6(e). The supervisor properly pick the candidate controller leading to better transient performance, and the safe adaptive controller achieves better performance than the two candidate controllers when applied individually.

Simulation 2

In this simulation, $\Delta_m(s) = 0$, $a = 1$, $b = 1$, and the reference signal and the input disturbance d_u are shown in Figures 3-7(a) and 3-7(b), respectively. The parameters involved in the adaptive law are chosen as $\Lambda_p(s) = s+10$, $\sigma = 10^{-5}$, $\Gamma = \text{diag}\{10, 10\}$ and the normalization term is $m^2 = 1 + 0.1\phi^T\phi + 0.01n_s$. The error signals ($e_1 = y_p - y_m$) for the two candidate controllers when applied individually are shown in Figure 3-6(c). The safe adaptive controller achieves better performance and its error and switching signals are shown in Figures 3-7(d) and Figure 3-7(e), respectively.

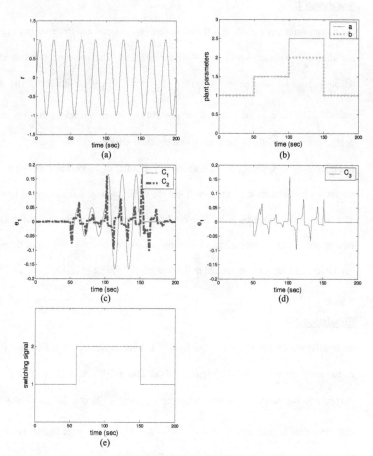

Figure 3-6: Results of Simulation 1.

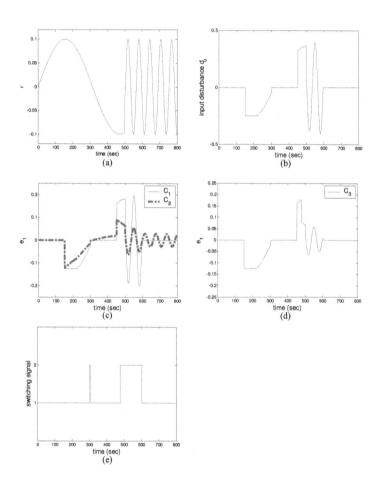

Figure 3-7: Results of Simulation 2.

82

Simulation 3

In this simulation, the modeling error is set as $\Delta_m(s) = \dfrac{0.1s}{0.1s+1}$. The reference signal,

input disturbance d_u and the true plant parameters are shown in Figures 3-8(a), 3-8(b) and

3-8(c), respectively. The parameters involved in the adaptive law are chosen as

$\Lambda_p(s) = s+10$, $\sigma = 10^{-5}$, $\Gamma = \text{diag}\{10, 10\}$ and the normalization term is $m^2 = 1 + 0.1\phi^T\phi$

$+ 0.01n_s$. The error signals ($e_1 = y_p - y_m$) for the two candidate controllers when applied

individually are shown in Figure 3-8(d). The supervisor properly switches between the

two candidate controllers and better performance has been achieved, as shown in Figure

3-8(e). The switching signal is shown in Figure 3-8(f).

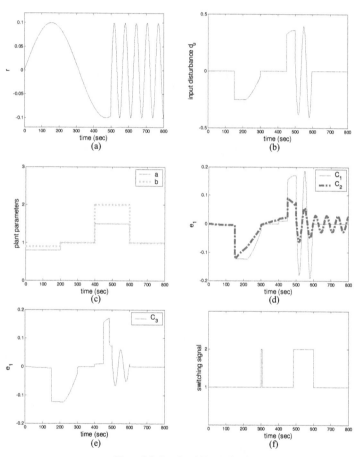

Figure 3-8: Results of Simulation 3.

84

3.6 Concluding Remarks

The safe adaptive controller investigated in this chapter is composed of a supervisor and two candidate controllers. One candidate controller is a LTI model reference controller that can always guarantee system stability, while the other is a model reference adaptive controller that may or may not guarantee system stability. The supervisory evaluates the performance of the two candidate controllers without using any plant model and activates the candidate controller leading to better transient response. It is shown that the system stability is guaranteed with the properly designed supervisor even if the adaptive candidate controller is destabilizing. The simulation results show that the adaptive candidate controller is activated in a conservative way, and the developed safe adaptive controller is able to achieve better performance than either of the candidate controllers. The safe adaptive control scheme investigate in this chapter will boost the applications of various identifier based adaptive controllers in the fields where their stability and performance are not trusted. It is currently under investigation whether the safe adaptive control scheme can modified and applied into the general trajectory tracking case in which the adaptive controllers are designed with the pole placement technique.

CHAPTER 4 : ADAPTIVE VEHICLE FOLLOWING

CONTROL DESIGN WITH VARIABLE TIME HEADWAY

4.1 Introduction

During the last decade considerable progress has been made in the area of Automated Highway Systems (AHS) both on the research and testing levels [9, 24, 64, 70, 80, 97]. Sensor technologies for ITS applications, control and communication protocols, field tests, and experiments in controlled environments have been carried out. Though dedicated highways with fully automated vehicles is a far in the future objective [30], the introduction of semi-automated vehicles, such as vehicles with Adaptive Cruise Control (ACC) system, also referred to as Intelligent Cruise Control (ICC) system, on current highways designed to operate with manually driven vehicles has already taken place in Japan and Europe and more recently in the United States [63]. The ACC system is an extension of the conventional Cruise Control (CC) system. In addition to the autonomous speed regulation provided by the conventional CC system, the ACC system provides the intelligent function that enables the ACC vehicle to adjust its speed automatically based on the traffic ahead. When the lane is clear, i.e. no object is detected by the forward-looking sensor installed on the vehicle, the ACC system works in the speed tracking mode and regulates the vehicle speed towards the desired speed set by the driver or the roadway controller (in the case of a future advanced traffic system). Otherwise, the ACC system works in the vehicle following mode and regulates the vehicle speed to maintain a desired spacing from the preceding vehicle or obstacle in the longitudinal direction.

The inter-vehicle spacing policy between traveling vehicles in the same lane, or equivalently the time headway used by the ACC system, is a critical parameter of an AHS system. It should be chosen as small as possible to increase the highway capacity but never to violate the safety constraint. In many studies related to ACC designs such as [28], the constant time headway policy was employed. Several vehicle following controllers were proposed and tested on real vehicles in [28]. It has been noticed that using variable time headway may lead to better traffic or vehicle following performance. In [91], a variable time headway was proposed for tightly vehicle following control, which depends on the speeds of the ACC vehicle and its preceding vehicle. Though no stability properties were established, the simulation results demonstrated that the vehicle following performance was improved by using such a variable time headway. Swaroop et al. [77] attempted to improve traffic flow performance involving ACC vehicles by using a variable time headway with traffic flow considerations, based on the hypothesis proposed by Greenshields [21]. In this case, the time headway is a strictly decreasing function of the traffic density, or equivalently, a strictly increasing function of the ACC vehicle's speed. It was demonstrated via simulations that the new spacing policy has a better effect on traffic flow than the constant time headway policy. Other implementations of variable time headway can be found in many literatures such as [12, 66, 83].

In [77] and [83], the vehicle following controller with variable time headway was designed using feedback linearization methodology based on a simplified second-order model. Though the design procedure is straightforward, this controller may generate high control efforts or even not be applicable for certain time headway such as that studied in [91], as we will show in the next section. In this chapter, we design vehicle following controller that can provide desired stability properties for a class of variable time headway including those in [12, 66, 77, 83, 91] as well as the constant time headway. Simulations are conducted to demonstrate the performance of the proposed controller.

In the ACC design, lateral control is the responsibility of the driver and driving in the ACC mode involves lanes with small curvatures. Therefore the lateral dynamics are considered to be decoupled from the longitudinal dynamics. In this chapter, we refer to the longitudinal vehicle model as the vehicle model. The rest of this chapter is organized as follows. In section 4.2 the validated nonlinear vehicle model is briefly discussed. In section 4.3, we present the simplified vehicle model for control design, the general form of variable time headway, and the proposed vehicle following controller. Stability analysis is also given in this section. In section 4.4, simulation results are presented to demonstrate the performance of the proposed vehicle following controller. The conclusions are given in section 4.5.

4.2 Vehicle Model

The longitudinal vehicle model used in this study is the same as that in [28], which is an experimentally validated nonlinear vehicle model. For completeness, we briefly review the model and detailed information can be found in [28].

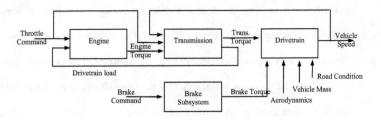

Figure 4-1: Block diagram of the vehicle model.

The nonlinear vehicle model can be described by a block diagram shown in Figure 4-1 [28]. The output of the engine subsystem is the engine torque that is a nonlinear function of the throttle angle and the drivetrain load. The transmission subsystem in Figure 4-1 is an automatic transmission with hydraulic toque coupling and four forward transmission gears. It transfers engine torque to the drivetrain depending on the vehicle speed and engine state. The drivetrain subsystem generates vehicle speed based on the received transmission torque and/or braking torque. The brake subsystem receives braking command and generates braking torque, and it behaves like a first order low pass filter with some time delay. It has been shown in [28] that this validate nonlinear vehicle model is good for the simulation purpose. It is used in our study for the simulations presented in section 4.4.

4.3 Vehicle Following Control Design

4.3.1 Simplified Vehicle Model for Control Design

The complexity of the model presented in section 4.2 makes it difficult to design the vehicle following controller. However, it has been shown that a simplified model with unknown parameters is quite good for the control design purpose [28, 91], which can be expressed as

$$\dot{v} = -a(v - v_d) + b(u - u_d) + d \qquad (4\text{-}1)$$

where v is the longitudinal vehicle speed, u is the throttle/braking command, v_d is the desired steady state speed, u_d is the corresponding steady state fuel command, d is the modeling uncertainty, and a and b are constant parameters that depend on the operating point. If there is no shift of gears, a and b are positive. In our analysis, we assume that a and b are always positive since any gear shift results in a short time transient activity. For a given vehicle, the relationship between v_d and u_d can be described by a smooth function of 1-1 mapping

$$v_d = f_u(u_d) \qquad (4\text{-}2)$$

In the vehicle following mode, the desired speed for the following vehicle is v_l, the speed of the preceding vehicle. Hence, the simplified vehicle model used for control design is described by (4-1) and (4-2) with v_d replaced by v_l. In our analysis, it is also assumed that d, $\dot{d}$, v_l and $\dot{v}_l$ are all bounded.

Remark 4-1: The simplified model given by (4-1) and (4-2) is only used for control design while the validated nonlinear model introduced in section 4.2 is used for simulations. In the control design, a and b are treated as unknown positive constant parameters since they changes when gears are shifted or other variables change. We use the adaptive control methodology to update the control gains on-line so that the desired performance can be achieved.

Remark 4-2: For control design and analysis, we assume that the relationship between v_d and u_d can described by the smooth function (4-2). It may be argued that the relationship is rather complicated and cannot be described by a smooth function in the practical situation since gear shifting and driving condition will affect the steady state speed for a given value of u_d. However, the discrepancy between v_d and $f_u(u_d)$ can always be merged into the model uncertainty term d and it won't change the boundness property of d and $\dot{d}$. Therefore, it is a reasonable assumption that the function (4-2) is smooth in the simplified control design model.

4.3.2 Control Objective and Constraints

In the vehicle following mode, the ACC system regulates the vehicle speed so that it follows the preceding vehicle by maintaining a desired inter-vehicle spacing. The control objective is to regulate the vehicle speed v to track the speed of the preceding vehicle v_l while maintaining the inter-vehicle spacing x_r as close to the desired spacing s_d as possible, as shown in Figure 4-2.

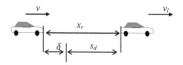

Figure 4-2: Diagram of the vehicle following mode.

With the time headway policy, the desired inter-vehicle spacing is given by

$$s_d = s_0 + hv \tag{4-3}$$

where s_0 is a fixed safety inter-vehicle spacing to avoid vehicle contact at low or zero speeds, v is the speed of the following vehicle, and h is the time headway. The control objective in the vehicle following mode can be expressed as

$$v_r \to 0, \delta \to 0 \text{ as } t \to \infty \tag{4-4}$$

where $v_r = v_l - v$ is the speed error and $\delta = x_r - s_d$ is the separation error. In practice, this control objective may not be met exactly when v_l varies considerably, or in the presence of sensor noise, modeling errors, delays and other imperfections. For safety reasons the time headway should be chosen large enough and the ACC system should be designed to be robust with respect to these imperfections so that δ remains non-negative most of the time, i.e. the ACC vehicle follows the preceding vehicle in a safe manner.

The following two constraints should also be satisfied:

C1. $a_{\min} \leq \dot{v} \leq a_{\max}$ where $a_{\min}$ and $a_{\max}$ are specified.

C2. The absolute value of jerk defined as $|\ddot{v}|$ should be small.

The above constraints are the result of driving comfort concerns and are established using human factor considerations [28].

4.3.3 Variable Time Headway

Most of the previous studies for vehicle following control considered constant spacing rules (h=0) [69] and constant time headway spacing rule (h is a positive constant) [28, 78]. Some studies have indicated using variable time headway may lead to better impacts on highway traffic. In [12], the spacing policy is chosen as

$$s_d = s_0 + h_1 v + h_2 v^2 \tag{4-5}$$

where h_1 and h_2 are two positive constants. The time headway incorporated in (4-5) can be expressed as

$$h = h_1 + h_2 v \tag{4-6}$$

This time headway increases with v. In practice, however, vehicle speed cannot exceed certain limit v_{max}. Hence the time headway in (4-6) is in fact the same as

$$h = \begin{cases} h_1 + h_2 v, & \text{if } v < v_{\max} \\ h_1 + h_2 v_{\max}, & \text{otherwise} \end{cases} \tag{4-7}$$

In [77, 83], the time headway is chosen based on the hypothesis proposed by Greenshields [21], and it can be written as

$$h = \frac{1}{k_{jam}\left(v_{free} - v\right)} \tag{4-8}$$

where k_{jam} is the traffic density corresponding to the congestion conditions and v_{free} is the free speed when the traffic density is low. The time headway in (4-8) is expressed differently from that in [83] because the spacing considered in [83] incorporates the vehicle length. Similar to the time headway in (4-6), the time headway in (4-8) can only be applicable to speeds lower than v_{max}. In [91], the time headway h proposed for tightly vehicle following control is given as

$$h = sat(h_0 - c_h v_r) \qquad (4\text{-}9)$$

where h_0 and c_h are positive constants to be designed, the saturation function sat($\bullet$) has an upper bound 1 and a lower bound 0. Though sat($\bullet$) is not analytical when v_r is equal to h_0/c_h or $(h_0-1)/c_h$, slight modification will change h to a smooth function of v and v_l.

In this chapter, we consider a general time headway that is a smooth function of v and v_l, i.e. $h(v, v_l)$. Let us define

$$H = \frac{\partial}{\partial v} s_d(v, v_l) \qquad (4\text{-}10a)$$

$$H_l = \frac{\partial}{\partial v_l} s_d(v, v_l) \qquad (4\text{-}10b)$$

This general time headway has the properties that $H \geq 0$ and H and H_l are bounded. With the practical consideration as given in (4-7), and the proper modification for the smoothness of h in (4-9), we can see the general time headway includes all the cases mentioned above. In particular, for the constant time headway, H is equal to the time headway h and H_l is zero.

4.3.4 Control Design

The simplified vehicle model described by (4-1) and (4-2) is used for the vehicle following control design. In [77, 83], a vehicle following controller using the variable time headway in (4-7) was proposed based on feedback linearization. The desired closed-loop system is described by

$$\dot{\delta} = -k\delta \qquad\qquad (4\text{-}11)$$

where k is a positive constant. If all the parameters in the simplified model are known, the vehicle following controller

$$u = \frac{1}{bH}\left(v_r + k\delta\right) - \frac{a}{b}v_r + u_d - \frac{d}{b} - \frac{H_l}{bH}\dot{v}_l \qquad\qquad (4\text{-}12)$$

can be used to achieve the desired closed-loop system in (4-11). The design procedure is straightforward, but the controller in (4-12) has some limitations. When H is very small, it may generate high control effort even for small separation and speed errors, which is undesired due to the control constraints presented in section 4.3.2. Furthermore, this controller cannot be implemented when H is equal to zero. Obviously H is zero in the constant spacing rule and may also be zero when the nonlinear time headway given in (4-9) is used. The vehicle following controller proposed this chapter, however, guarantees global stability for any variable time headway that has bounded H and H_l with $H \geq 0$. Furthermore, it always generates smooth control effort when the control parameters are properly chosen.

One important issue associated with vehicle following control is string stability. String stability in a vehicle string implies that any nonzero speed and separation error of an individual vehicle does not get amplified as it propagates upstream [10, 76]. Here we consider only L_2 string stability [10]. In the string stability analysis, we assume that the lead vehicle slowly varies its speed around the nominal value v_{l0}, and the vehicle following controller can sufficiently keep v_r and δ close to zero.

Lemma 4-1: Consider the system in (4-1) and (4-2), with the following controller

$$u = k_1^* v_r + k_2^* \Delta(\delta, t) + k_3^* \tag{4-13}$$

where $k_1^* = (a_m - a)/b$, $k_2^* = a_m/b$, $k_3^* = u_d - d/b$, Δ is a design time varying function of δ satisfying

$$k_l \delta \leq \Delta(\delta, t) \leq k_u \delta \tag{4-14}$$

and a_m, k_l and k_u are positive design constants.

(i) All the signals in the closed-loop system are bounded if a_m, k_l and k_u are designed such that there exists a positive constant p_1 satisfying

$$a_m p_1 > 1 \tag{4-15a}$$

$$a_m p_1^2 (k_u - k_l)^2 - 4k_u (a_m p_1 - 1) < 0 \tag{4-15b}$$

$$a_m + a_m k_l p_1 - k_u \geq 0 \tag{4-15c}$$

$$\frac{4 p_1 k_l}{a_m + a_m k_l p_1 - k_l} > \sup H \tag{4-15d}$$

where $\sup H$ is the supremum of H. Furthermore, if v_l is a constant, then the control objective in (4-4) is achieved, i.e. $v_r, \delta \to 0$ as $t \to \infty$.

(ii) If all the vehicles in the vehicle string are of the same type and equipped with (4-13) and Δ is a time invariant and differentiable function of δ, then *local L_2* string stability is guaranteed provided that the control parameters are chosen such that

$$a_m \Delta_\delta (0) \left(H^2 - H_l^2 \right) + 2 a_m \left(H + H_l \right) - 2 > 0 \qquad (4\text{-}16)$$

where $\Delta_\delta = \partial \Delta / \partial \delta$.

Proof: Let $\Delta(\delta, t) = k\delta$, where k is a time varying function of δ and satisfies $k_l \le k \le k_u$. Using (4-13) in (4-1) and (4-2), the closed-loop system is

$$\dot{v} = a_m \left(v_r + k\delta \right) \qquad (4\text{-}17)$$

(i) Denote $x_1 = v_r$ and $x_2 = \delta$, then

$$\begin{cases} \dot{x}_1 = -a_m \left(x_1 + kx_2 \right) + u_1 \\ \dot{x}_2 = \left(1 - a_m H \right) x_1 - a_m k H x_2 - H_l u_1 \end{cases} \qquad (4\text{-}18)$$

where $u_1 = \dot{v}_l$, H and H_l are bounded with $H \ge 0$ (recall the assumptions in section 4.3.3).

Consider the following candidate Lyapunov function

$$V_a = \tfrac{1}{2} x^T P x \qquad (4\text{-}19)$$

where $P = \begin{bmatrix} p_1 & 1 \\ 1 & p_2 \end{bmatrix}$ a positive definite matrix, and p_1 and p_2 are positive constants.

Hence,

$$\begin{aligned} \dot{V}_a &= p_1 x_1 \dot{x}_1 + p_2 x_2 \dot{x}_2 + x_1 \dot{x}_2 + \dot{x}_1 x_2 \\ &= p_1 x_1 \left[-a_m \left(x_1 + kx_2 \right) + u_1 \right] + p_2 x_2 \left[\left(1 - a_m H \right) x_1 - a_m k H x_2 - H_l u_1 \right] \\ &\quad + x_1 \left[\left(1 - a_m H \right) x_1 - a_m k H x_2 - H_l u_1 \right] + x_2 \left[-a_m \left(x_1 + kx_2 \right) + u_1 \right] \\ &= -x_1^2 \left(a_m p_1 - 1 + a_m H \right) - x_2^2 \left(a_m k + a_m k H p_2 \right) \\ &\quad - x_1 x_2 \left(a_m k p_1 + a_m - p_2 + a_m H p_2 + a_m k H \right) + u_1 \left(p_1 x_1 - H_l x_1 + x_2 - H_l p_2 x_2 \right) \end{aligned} \qquad (4\text{-}20)$$

We choose

$$p_2 = a_m k_l p_1 + a_m \qquad (4\text{-}21)$$

When (4-15a) holds, the coefficient of x_1^2 in (4-20) is negative and (4-21) guarantees that P is positive definite. With (4-21), (4-20) can be rewritten as

$$\dot{V}_a = -x_1^2\left(a_m p_1 - 1 + a_m H\right) - x_2^2\left[a_m k + a_m^2 kH(k_l p_1 + 1)\right] \\ - x_1 x_2\left(a_m kp_1 - a_m k_l p_1 + a_m^2 H + a_m^2 k_l Hp_1 + a_m kH\right) + u_1\left(w_1 x_1 + w_2 x_2\right) \qquad (4\text{-}22)$$

where $w_1 = p_1 - H_l$ and $w_2 = 1 - H_l(a_m k_l p_1 + a_m)$. Suppose u_1 is zero. Then $\dot{V}_a$ is negative definite if

$$\left(a_m kp_1 - a_m k_l p_1 + a_m^2 H + a_m^2 k_l Hp_1 + a_m kH\right)^2 \\ < 4\left(a_m p_1 - 1 + a_m H\right)\left[a_m k + a_m^2 kH(k_l p_1 + 1)\right] \qquad (4\text{-}23)$$

always holds. (4-23) can be rewritten as

$$a_m\left(a_m + a_m k_l p_1 - k\right)^2 H^2 - 2a_m p_1\left(k + k_l\right)\left(a_m + a_m k_l p_1 - k\right)H \\ + a_m p_1^2\left(k - k_l\right)^2 - 4k\left(a_m p_1 - 1\right) < 0 \qquad (4\text{-}24)$$

One necessary condition for (4-24) to be true for all $H \geq 0$ is that

$$a_m p_1^2\left(k - k_l\right)^2 - 4k\left(a_m p_1 - 1\right) < 0 \qquad (4\text{-}25)$$

for all $k \in [k_l, k_u]$. This condition is equivalent to that (4-15b) is true. When (4-15a) and (4-15b) hold, one sufficient condition for (4-24) to hold is that (4-15c) and (4-15d) hold. Now we have shown that if (4-15a) - (4-15d) hold and u_1 is zero then $\dot{V}_a$ is negative definite. Since w_1 and w_2 are bounded, it is easy to show that V_a is bounded, and then all the signals in the closed-loop system are bounded. Furthermore, if v_l is a constant, i.e. u_1

is zero, it can be verified that $x_1, x_2 \in L_2 \cap L_\infty$, $\dot{x}_1, \dot{x}_2 \in L_\infty$. It follows from Barbalat's Lemma [27] that $x_1, x_2 \to 0$ as $t \to \infty$, i.e., the control objective in (4-4) is achieved.

(ii) With linearization of (4-17), the transfer function from v_l to v is given as

$$G_v(s) = \frac{v}{v_l} = \frac{\left(a_m - a_m \Delta_\delta(0) H_l(v_{l0})\right)s + a_m \Delta_\delta(0)}{s^2 + \left[a_m \Delta_\delta(0) H(v_{l0}) + a_m\right]s + a_m \Delta_\delta(0)} \tag{4-26}$$

If we assume all the vehicles in the same lane are of the same type and equipped with the same controller, the linearized vehicle string is L_2 string stable [10] if

$$\left|G_v(j\omega)\right| < 1, \forall \omega > 0 \tag{4-27}$$

which is equivalent to (4-16).

$\square$

Remark 4-3: Here we can only show the local string stability since the controller in (4-13) results in a nonlinear string, which makes the analysis hard. The practical string stability is even harder to analyze due to large time delays and modeling uncertainties in the system [43].

Remark 4-4: The condition for L_2 string stability involving constant time headway policy presented in [9] agrees with (4-16) if we set $\Delta_\delta = k$, $H = h$ and $H_l = 0$. It is interesting to notice that if $H_l=0$ and H is very small, then a_m and $\Delta_\delta(0)$ should be chosen large to guarantee local string stability for all speeds. However, for reasonable choices of time headway, H should be small when v is small and relatively large when v is large [15].

Hence we can avoid high control gain by sacrificing string stability at low speeds (in which cases H is small).

Remark 4-5: From (4-26), it can also be found out why the time headway chosen as (4-9) could lead to better string stability performance [10]. If the proposed controller can efficiently keep v_r small, then H is much larger than H_l at high speeds. Hence (4-26) implies that the time headway in (4-9) can achieve better string stability performance than a constant time headway h_0.

The controller in (4-13) cannot be implemented because a, b and d are unknown parameters which may change with vehicle speed and other conditions. However, we can estimate k_i^* (i=1,2,3) on-line and use their estimates in the control law. In the next Lemma, we show that with proper update laws for k_i, the control law (4-13) where the k_i^* (i=1,2,3) are replaced with their on line estimates meets the control objective.

Lemma 4-2: Consider the system in (4-1) and (4-2), with the control law

$$u = k_1 v_r + k_2 \Delta(\delta, t) + k_3 \qquad (4\text{-}28)$$

where k_i is the estimate of k_i^* (defined in Lemma 4-1) with initial condition k_{i0} (i=1,2,3), generated by the adaptive laws

$$\begin{cases} \dot{k}_1 = \mathrm{Proj}\{\gamma_1 x_1 [(p_1 x_1 + x_2) + (x_1 + a_m k_l p_1 x_2 + a_m x_2)H]\} \\ \dot{k}_2 = \mathrm{Proj}\{\gamma_2 k x_2 [(p_1 x_1 + x_2) + (x_1 + a_m k_l p_1 x_2 + a_m x_2)H]\} \\ \dot{k}_3 = \mathrm{Proj}\{\gamma_3 [(p_1 x_1 + x_2) + (x_1 + a_m k_l p_1 x_2 + a_m x_2)H]\} \end{cases} \qquad (4\text{-}29)$$

where a_m, p_1, γ_1, γ_2, and γ_3 are positive design parameters, Proj$\{\bullet\}$ is the projection function keeping k_i within the intervals $[k_{il}, k_{iu}]$ (i=1,2,3). k_{il} and k_{iu} are chosen such that $k_i^* \in [k_{il}, k_{iu}]$. If we choose the parameters a_m, k_l, k_u and p_1 such that (4-15a) - (4-15d) hold, then all the signals in the closed-loop system are bounded. Furthermore, if v_l and d are constants, then the control objective in (4-4) is achieved, i.e. $v_r, \delta \to 0$ as $t \to \infty$.

Proof: With the proposed control law, the closed-loop system becomes

$$\dot{v} = a_m(v_r + k\delta) + b\tilde{k}_1 v_r + b\tilde{k}_2 \Delta(\delta, t) + b\tilde{k}_3 \tag{4-30}$$

where $\tilde{k}_i = k_i - k_i^*$ (i=1,2,3). We rewrite $\Delta(\delta, t)$ as $k\delta$ and denote $x_1 = v_r$ and $x_2 = \delta$. Now we have

$$\begin{cases} \dot{x}_1 = -a_m(x_1 + kx_2) - b\tilde{k}_1 x_1 - bk\tilde{k}_2 x_2 - b\tilde{k}_3 + u_1 \\ \dot{x}_2 = (1 - a_m H)x_1 - a_m kHx_2 - bH\tilde{k}_1 x_1 - bHk\tilde{k}_2 x_2 - bH\tilde{k}_3 - H_l u_1 \end{cases} \tag{4-31}$$

where $u_1 = \dot{v}_l$. Consider the following Lyapunov function

$$V = V_a + \sum_{i=1}^{3} \frac{b}{2\gamma_i} \tilde{k}_i^2 \tag{4-32}$$

where V_a is the same as in (4-20). It is easy to verify by using the adaptive laws (4-29) and the knowledge of $k_i^* \in [k_{il}, k_{iu}]$ that

$$\dot{V} \leq \dot{V}_a - \frac{b}{\gamma_3} \tilde{k}_3 \dot{k}_3^* \tag{4-33}$$

where $\dot{V}_a$ is given in (4-23) and

$$\dot{k}_3^* = \dot{u}_d - \frac{d}{b} \tag{4-34}$$

101

Since $\dot{d}$, $\dot{v}_l$ and the derivative of the function in (4-2) are assumed to be bounded, it follows that $\dot{k}_3^*$ is bounded. It is easy to show that if all the conditions in (4-15a) - (4-15d) are satisfied then $\dot{V}$ is negative when x_1 or x_2 or both are large. This implies that V is bounded, and therefore all the signals in the closed-loop system are bounded.

When v_l and d are constants, $\dot{V} < 0$ when either x_1 or x_2 is nonzero. It is easy to verify that $x_1, x_2 \in L_2 \cap L_\infty$ and $\dot{x}_1, \dot{x}_2 \in L_\infty$. It follows from Barbalat's Lemma that $x_1, x_2 \to 0$ as $t \to \infty$, i.e., the control objective in (4-4) is achieved .

$\square$

This adaptive controller provided by Lemma 4-2 can be used for any variable time headway with the properties that $H \geq 0$ and H and H_l are bounded. In [91], an adaptive controller was proposed with the nonlinear time headway in (4-9), and the gain k (i.e. Δ/δ) was chosen as

$$k = c_k + (k_0 - c_k)e^{-\sigma\delta^2} \tag{4-35}$$

where k_0, c_k, and σ are positive constants to be designed (with $c_k < k_0$). k was designed in this way to eliminate the adverse effect of large separation error. Even though it was shown in [91] that such choices for h and k could lead to good platoon performance, the system stability was not established. It can be observed that k in (4-35) is bounded by c_k and k_0 all the time. Hence Lemma 4-2 points out that if we choose the control parameters properly, the adaptive controller in (4-28) with the update law in (4-29) makes the closed-loop system stable with h and k as chosen in [91]. Simulation results will be presented in section 4.4 to demonstrate that the proposed controller can guarantee stability and achieve the desired performance.

Since we have the flexibility to choose $\Delta(\delta, t)$, we can set $\Delta(\delta, t) = k\delta$ where k is a positive constant. Hence we have the following lemma, which is a special case of Lemma 4-2 with the fact that $k_l = k = k_u$. The proof follows the same steps as the proof for Lemma 4-3, and is omitted.

Lemma 4-3: Consider the system in (4-1) and (4-2), with the control law

$$u = k_1 v_r + k_2 \delta + k_3 \qquad (4\text{-}36)$$

where k_i (with proper initial condition k_{i0} (i=1,2,3)) is generated according to the adaptive laws:

$$\begin{cases} \dot{k}_1 = \mathrm{Proj}\{\gamma_1 x_1 [(p_1 x_1 + x_2) + (x_1 + a_m k p_1 x_2 + a_m x_2)H]\} \\ \dot{k}_2 = \mathrm{Proj}\{\gamma_2 x_2 [(p_1 x_1 + x_2) + (x_1 + a_m k p_1 x_2 + a_m x_2)H]\} \\ \dot{k}_3 = \mathrm{Proj}\{\gamma_3 [(p_1 x_1 + x_2) + (x_1 + a_m k p_1 x_2 + a_m x_2)H]\} \end{cases} \qquad (4\text{-}37)$$

where a_m, p_1, γ_1, γ_2, and γ_3 are positive design parameters, and $\mathrm{Proj}\{\bullet\}$ is defined in Lemma 4-2. All the signals in the closed-loop system are bounded if the design parameters are chosen such that

$$a_m p_1 > 1 \qquad (4\text{-}38a)$$

$$\frac{4 p_1 k}{a_m + a_m k p_1 - k} > \sup H \qquad (4\text{-}38b)$$

Furthermore, if v_l and d are constants, then the control objective in (4-4) is achieved, i.e. $v_r, \delta \to 0$ as $t \to \infty$.

103

The ACC system should never activate the fuel and brake systems at the same time, and it should also avoid unnecessary switching between the two sub-systems. Hence the following switching rules are incorporated in the vehicle following mode:

S1. If the separation distance x_r is larger than x_{max} ($x_{max}>0$ is a design constant), then the fuel system is on.

S2. If the separation distance x_r is smaller than x_{min} ($x_{min}>0$ is a design constant), then the brake system is on.

S3. If $x_{max} \leq x_r \leq x_{max}$, then the fuel system is on when $u>0$, while the brake system is on when $u<-u_0$ ($u_0>0$ is a design constant). When $-u_0 \leq u \leq 0$, the brake system is inactive and the fuel system is operating as in idle speed.

There are several other practical issues associated with the application of the controller (4-28) or (4-36). To guarantee that the constraints **C1**, **C2** are not violated, we should avoid the generation of high or fast varying control signals. Though the controller in (4-28) or (4-36) is proposed without high gains, such high or fast varying control signals can still be generated if the lead vehicle accelerates rapidly or changes lanes creating a large spacing error or the ACC vehicle switches to a new target with large initial separation or speed error. To eliminate the adverse effect of large separation error, the control parameter k shown in (4-35) is proposed in [91]. When a constant k is to be used, the function $sat(\delta)$ defined as

$$sat(\delta) = \begin{cases} e_{max}, & \text{if } \delta > e_{max} \\ e_{min}, & \text{if } \delta > e_{min} \\ \delta, & \text{otherwise} \end{cases} \tag{4-39}$$

can used instead of δ in order to deal with large separation error [28]. To eliminate the adverse effect of fast varying v_l, the nonlinear filter shown in Figure 4-3 was used in [28] to smooth the speed trajectory of the lead vehicle. The filtered speed trajectory $\hat{v}_l$, instead of v, is then used by the controller. Furthermore, a low pass filter is placed before the throttle actuator so that fast varying commands will not be sent to the throttle system. The modifications in [28] are adopted in our vehicle following controller and evaluated using simulations.

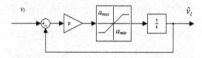

Figure 4-3: Nonlinear filter used to smooth v_l.

4.4 Simulations

In this section, we present the simulation results that demonstrate the performance of the vehicle following controller given by (4-28) with the nonlinear spacing rule using h in (4-9) and k in (4-35). As a special case of (4-28), the controller (4-36) is also tested for a string of 5 ACC vehicles using the time headway in (4-7). The vehicle model used for the simulations is the validated nonlinear model presented in section 4.2.

4.4.1 Simulation 1

In this simulation, the controller given by (4-28) with the update law in (4-29) is tested using h in (4-9) and k in (4-35). The control parameters are chosen as

$s_0 = 4.5\text{m}, h_0 = 0.5, c_h = 0.1, k_0 = 0.5, c_k = 0.2,$

$a_m = 0.5, p_1 = 10,$

$a_{max} = 1.0\text{m/s}^2, a_{min} = -2.0\text{m/s}^2, p = 10$

$k_{10} = 6, k_{1u} = 12, k_{1l} = 4, \gamma_1 = 0.1,$

$k_{20} = 2, k_{2u} = 3, k_{2l} = 0.5, \gamma_2 = 0.05,$

$k_{30} = 0, k_{3u} = 30, k_{3l} = -30, \gamma_3 = 0.02$

It can be verified that the control parameters are chosen such that (4-15a)-(4-15d) are satisfied when the lead vehicle operates under the maximum speed 30m/s.

Two vehicles are used to evaluate the vehicle following properties of the proposed controller. The following vehicle is equipped with the proposed controller given by (4-28) and (4-29). At time zero, the two vehicles have zero speed and are separated with a distance of s_0. From $t = 0$s to $t = 20$s, the lead vehicle increases its speed with a constant acceleration 0.8m/s², and then cruises at 16m/s. From $t = 50$s to $t = 53$s, the lead vehicle increases its speed with a constant acceleration 2.0m/s², and then cruises at 22m/s. From $t = 90$s to $t = 100$s, the lead vehicle increases its speed with a constant acceleration 0.6m/s², and then cruises at 28m/s. From $t = 140$s to $t = 144$s, the lead vehicle decreases its speed deceleration -2.0m/s^2, and then cruises at 20m/s.

The speed of the lead vehicle is presented as the dotted line in Figure 4-4(a). The speed, acceleration, speed error, separation error, throttle angle and brake pressure responses of the following vehicle are presented in Figures 4-4(a) to 4-4(f), respectively. As we can see, when the acceleration of the lead vehicle is not too large (0.8 or 0.6m/s^2), the throttle controller regulates the fuel system smoothly and the ACC vehicle follows the lead vehicle with small speed and separation errors. These errors are regulated towards zero when the lead vehicle reaches a constant speed. When the acceleration of the lead vehicle is large (2.0m/s^2) for a short time, the following vehicle increases its speed in a smooth and comfortable way. The transient speed and separation errors are large due to the high acceleration of the lead vehicle. However, the errors are regulated towards zero as soon as the lead vehicle reaches a constant speed. When the lead vehicle decreases its speed rapidly, the brake system on the ACC vehicle is active and the brake pressure is shown in Figure 4-4(f). From the acceleration and separation error responses, we can see that the ACC system regulates the vehicle speed in a comfortable and safe way.

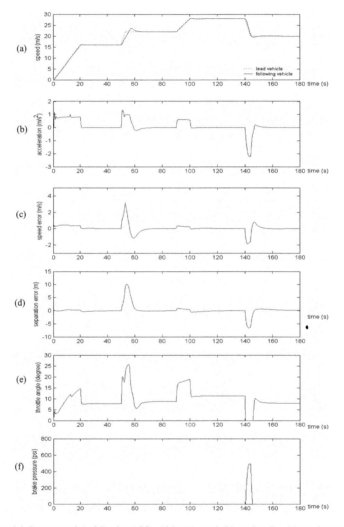

Figure 4-4: Responses of the following ACC vehicle: (a) speed, (b) acceleration, (c) speed error, (d) separation error, (e) throttle angle and (f) brake pressure in Simulation 1.

4.4.2 Simulation 2

In this simulation, the controller given by (4-36) with the update law in (4-37) is tested using the h in (4-7) and a constant k. The control parameters are chosen as

$s_0 = 4.5$m, $h_1 = 0.5$, $h_2 = 0.016$, $k = 0.2$, $a_m = 2$, $p_1 = 10$

$a_{max} = 1.0$m/s^2, $a_{min} = -2.0$m/s^2, $p = 10$,

$e_{max} = 10$m, $e_{min} = -30$m,

$k_{10} = 8$, $k_{1u} = 14$, $k_{1l} = 4$, $\gamma_1 = 0.28$,

$k_{20} = 2$, $k_{2u} = 3$, $k_{2l} = 0.5$, $\gamma_2 = 0.1$,

$k_{30} = 0$, $k_{3u} = 30$, $k_{3l} = -30$, $\gamma_3 = 0.08$

The control parameters are chosen such that (4-16), (4-38a) and (4-38b) are satisfied. This simulation is used to demonstrate that the proposed vehicle following controller in (4-36) provides not only stability for each individual following vehicle but also string stability for the vehicle string.

In this simulation five vehicles are simulated in a vehicle following scenario. The lead vehicle generates the speed trajectory shown as a dotted line in Figure 4-5(a). The four following vehicles are equipped with the proposed controller in (4-36) with the update law in (4-37). The speed, speed error, and separation error responses of the following ACC vehicles are shown in Figures 4-5(a) to 4-5(c), respectively. As we can see the speed and separation errors are attenuated within the vehicle string when the brake system is not activated. When the lead vehicle begins to decelerate rapidly, the speed and separation errors are not attenuated within the vehicle string. This is because the switching logic introduces a hysteresis effect in the ACC system. However, the errors are still propagated in a satisfactory manner. All these results indicate the proposed controller can provide desired stability properties.

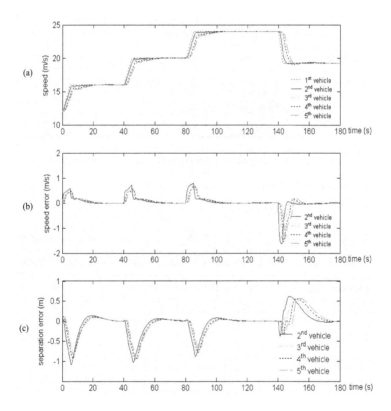

Figure 4-5: Responses of the following ACC vehicles: (a) speed, (b) speed error and (c) separation error in Simulation 2.

4.5 Conclusions

In this chapter, we design an adaptive vehicle following control system with a general variable time headway. It is show that the proposed controller guarantees system stability, and that the control objective can be achieved when the lead vehicle operates at a constant speed. The analysis can be extended to any time headway with bounded partial derivatives with respect to the speeds of lead and following vehicles. Simulations using validated nonlinear vehicle model demonstrate our analytical results, and the designed control system provides desired stability performance.

In contrast to the feedback linearization control design, the controller proposed in this chapter will not lead to high control effort even when H (the partial derivative h with respect to v) is small. It can better take care of the control constraints for driving comfort. This improvement is achieved by sacrificing string stability when H is small. However, the proposed controller is still proper since H is relatively large at high speeds for a reasonable choice of time headway. Furthermore, it can always provide stability for each individual following vehicle when H is zero.

CHAPTER 5 : ADAPTIVE VEHICLE FOLLOWING

CONTROL DESIGN WITH IMPROVED PERFORMANCE

5.1 Introduction

It is envisioned that automation in heavy trucks is more likely to be introduced than in passenger vehicles due to cost as well as human factor considerations. Extensive studies have been done on longitudinal vehicle following control, and in particular on adaptive cruise control (ACC), also referred to as intelligent cruise control (ICC). DaimlerChrysler has already developed automatic vehicle following control systems for heavy trucks, referred to as "electronic draw bar" system, based on electronic sensors used for things such as brake-by-wire and collision avoidance systems {Parker, July 12, 1999}. Others include the Eaton-VORAD Collision Avoidance System which allows a truck to perform automatic vehicle following maintaining a safe time headway in traffic [33]. At PATH, there have been several research efforts on truck automation [23, 89-92]. The environmental performance of these automatic control systems as they begin to penetrate into the transportation system is very important. The expected benefits and positive effects once validated by analysis and experiments will help accelerate the deployment of automated heavy trucks as well as other intelligent vehicles and concepts of Automated Highway Systems (AHS). In [11, 26] it has been shown that the smooth response of the ACC vehicles, designed for passenger comfort, significantly reduces fuel consumption and levels of pollutants during the presence of traffic disturbances. Studies on the potential impact of Advanced Public Transportation Systems (APTS) on air quality and fuel economy, concluded that transit buses produce less hydrocarbon (HC) and carbon monoxide (CO) emissions than autos on a passenger-mile basis [29].

In this chapter, we evaluate the performance of different vehicle following controllers currently available for heavy trucks using microscopic simulations. Our simulation results indicate that the performance of a vehicle following controller is mainly determined by the spacing policy employed rather than by its form. We therefore focused on proportional-integral-derivative (PID) type vehicle following controllers for different spacing policies. A new vehicle following controller is designed to provide better performance on the microscopic level with beneficial effects on fuel economy and pollution. The validated emission models for passenger vehicles and trucks developed in [8, 67] are used to evaluate the impact of trucks on emissions and fuel economy. Experiments with actual vehicles are used to demonstrate the operation of the longitudinal controllers in real time and to validate our simulation models.

The sluggish dynamics of trucks whether manual or ACC due to limited acceleration and speed capabilities make their response to disturbances caused by lead passenger vehicles smooth. Vehicles following the truck are therefore presented with a smoother speed trajectory to track. This filtering effect of trucks is shown to have beneficial effects on fuel economy and pollution. The quantity of the fuel and emission benefits depends very much on the level of the disturbance and scenario of maneuvers. If the response of the truck is too sluggish relative to the speed of the lead vehicle, then a

113

large inter-vehicle spacing may be created inviting cut-ins from neighboring lanes. These cut-ins create additional disturbances with negative effects on fuel economy and emissions. Situations can be easily constructed where the benefits obtained due to the filtering effect of trucks are eliminated due to disturbances caused by possible cut-ins. Furthermore cut-ins are shown to increase travel time for the vehicles in the original traffic stream.

The rest of this chapter is organized as follows. In section 5.2 the simulation models are presented. In section 5.3, we analyze a PID type vehicle following controller for different spacing policies and design a new vehicle following controller. In section 5.4 microscopic simulation results are used to demonstrate that the new controller can achieve better performance in the presence of traffic disturbances. In section 5.5, the simulation results are used to investigate the environmental performance of the new controller. In section 5.6, experimental and simulation data are presented and used to validate the models used in the simulations. The conclusions are presented in section 5.7.

5.2 Vehicle Models

5.2.1 Longitudinal Dynamics of Heavy Trucks

In this study, the longitudinal truck model used for simulations is the same as in [91, 92]. It is proposed in [89, 90] and experimentally validated in section 5.6. This simulation model consists of a turbocharged diesel engine, an automatic transmission, a longitudinal drivetrain, and a brake subsystem. It can be characterized by a set of differential

114

equations, algebraic relations and look-up tables. The dominant state in this model is the one associated with the longitudinal speed v, which is determined by

$$\dot{v} = \frac{F_t - F_a - F_r}{m} \qquad (5\text{-}1)$$

where m is the vehicle mass, F_t is the tractive tire force, F_a is the aerodynamic drag force, and F_r is the rolling friction force. In (5-1), F_a is equal to $c_a v^2$, where c_a is the aerodynamic drag coefficient, and F_r is equal to $c_r mg/h_w$, where c_r is the rolling friction coefficient, h_w is the radius of the front wheels and g is the gravity constant. The brake/fuel commands are incorporated in the differential and algebraic equations that determine the tire force F_t. The complete nonlinear model is rather complicated and can be found in [89, 90]. As in every control design the proposed controller in this chapter is based on a simplified truck model but it is tested and evaluated using the complex nonlinear validated truck model.

5.2.2 Human driver model for passenger vehicles

The Pipes' vehicle following model introduced in [62] is chosen to simulate manually driven passenger vehicles, since it is found to be the most appropriate human driver model for the type of simulations and experiments we perform in the sense that it accurately models the vehicle following behavior of human drivers in the presence of traffic disturbances [11]. The Pipes' model is described as

$$a_i(t) = K[v_{i-1}(t - \tau) - v_i(t - \tau)] \qquad (5\text{-}2)$$

115

where v_{i-1} and v_i are the speeds of the i–1th (leading the i-th vehicle) and ith vehicles, respectively, a_i is the acceleration of the ith vehicle, K is the sensitivity factor and τ is the reaction delay. In the simulations, we use τ=1.5sec and K=0.37sec^{-1} [11].

5.2.3 Human driver model for heavy trucks

In the human driver model proposed by Bando et al. [7], the driver controls the acceleration in order to maintain an optimal safe speed according to the following distance to the preceding vehicle. The human driver model for heavy trucks used in our study is a modified version of the Bando's model and it is expressed as

$$a_i(t) = \begin{cases} A_{\min}, & \text{if } K_t(V(x_{i-1}(t-\tau_t) - x_i(t-\tau_t)) - v_i(t)) < A_{\min} \\ A_{\max}, & \text{if } K_t(V(x_{i-1}(t-\tau_t) - x_i(t-\tau_t)) - v_i(t)) > A_{\max} \\ K_t(V(x_{i-1}(t-\tau_t) - x_i(t-\tau_t)) - v_i), & \text{otherwise} \end{cases} \qquad (5-3)$$

where τ_t is a time delay, K_t is the sensitivity factor for heavy trucks, x_{i-1} and x_i are the positions of the i–1th vehicle and the ith vehicle in a vehicle string, $A_{\min}$ and $A_{\max}$ are the minimum acceleration (i.e. maximum deceleration) and maximum acceleration, and $V(\cdot)$ is the desired speed as a function of the vehicle separation distance. In the simulations, we assume that the truck driver attempts to maintain a constant time headway during driving, and the desired speed is given by

$$V(\Delta x_i) = \max\left\{0, \frac{\Delta x_i - s_{t0}}{h_{tm}}\right\} \qquad (5-4)$$

where $\Delta x_i = x_{i-1} - x_i$ is the inter-vehicle spacing, s_{t0} is the inter-vehicle spacing corresponding to the jam density, h_{tm} is the time headway used by the truck driver. If there is no acceleration limit, K_t is infinitely large and the desired speed in (5-4) is not

116

restricted to be positive. In this case the modified Bando's model given by (5-3) and (5-4) becomes identical to Pipes' model. In the simulations, we take $K_f=0.8s^{-1}$, which is recommended in [45]. The time delay τ_t is taken as 1.0 second, which is shorter than the delay in the Pipes' model. The reason is that since the truck driver sits high up and can view traffic far ahead he/she is able to make deceleration/acceleration decisions faster than drivers in passenger vehicles who can usually view only the vehicle in front of them. The time headway h_{tm} is set as 3.0 seconds, which is larger than that used in the Pipes' model for passenger vehicles. The reason is that the truck has lower braking acceleration capabilities than passenger vehicles and truck drivers need to maintain longer headway for safety considerations. We use $s_{t0}=6.0$m to account for the inter-vehicle spacing at zero or very low speeds.

Remark 5-1: Note that in (5-3) and (5-4) we assume that trucks share the same parameters and functions for simplicity purposes. The function V is the same for all the trucks, its argument however is different as it depends on the relative spacing of the individual truck.

5.3 Vehicle Following Control Design

5.3.1 Control Objective and Constraints

In the vehicle following control, the ACC system regulates the vehicle speed v towards the speed of the lead vehicle v_l and keeps the inter-vehicle spacing x_r close to the desired spacing s_d. With the time headway policy, s_d is given by

$$s_d = s_0 + hv \tag{5-5}$$

where s_0 is a fixed safety inter-vehicle spacing to avoid vehicle contact at low or zero speeds and h is the time headway. The control objective in the vehicle following mode can be expressed as

$$v_r \to 0, \delta \to 0 \text{ as } t \to 0 \tag{5-6}$$

where $v_r = v_l - v$ is the relative speed and $\delta = x_r - s_d$ is the separation error. This control objective should be achieved under the following two constraints [28]:

C1. $a_{min} \leq \dot{v} \leq a_{max}$, where a_{min} and a_{max} are specified.

C2. The absolute value of jerk defined as $|\ddot{v}|$ should be small.

The first constraint arises from the inability of the truck to generate high accelerations and from driver comfort and safety considerations. The second constraint is for driver comfort. While constraint C2 is more crucial in passenger vehicles it would be desirable to have it in trucks too.

In our analysis, we assume that v_l and its derivative are continuous and bounded signals as this is the case in practice.

118

5.3.2 Simplified Vehicle Model for Control Design

The simplified longitudinal model used for vehicle following control design is [28, 91, 92]

$$\dot{v} = -a(v - v_d) + b(u - u_d) + d \tag{5-7}$$

where v_d is the desired steady state speed, u_d is the corresponding steady state fuel command, d is the modeling uncertainty, and a and b are unknown constant parameters that depend on the operating point, i.e. the steady state values of the vehicle speed and load torque. If there is no shift of gears, a and b are positive. In our analysis, we always assume a and b are positive since gear shift is a transient activity. We also assume that d and $\dot{d}$ are bounded. For a given vehicle, the relationship between v_d and u_d can be described by a look-up table, or by a smooth function of 1-1 mapping

$$v_d = f_u(u_d) \tag{5-8}$$

In the vehicle following mode, the desired steady state speed is the speed of the lead vehicle. Hence, the simplified truck model used for control design is described by (5-7) and (5-8) with $v_d = v_l$.

5.3.3 Control Design

In most studies related to vehicle following control such as [28], the time headway is chosen as a positive constant. A number of vehicle following controllers can be designed based on the simplified vehicle model, and simulations or experiments demonstrate that they all work well when the control parameters are properly chosen [28, 91, 92]. In this study, we use a PID type controller

119

$$u = k_p\left(v_r + k\delta\right) + k_i\frac{1}{s}\left(v_r + k\delta\right) + k_d\frac{s}{\frac{1}{N}s+1}\left(v_r + k\delta\right) \qquad (5\text{-}9)$$

where s is the Laplace operator, k, k_p, k_i, k_d and N are positive control parameters to be chosen. With linear control analysis, it can be shown that the closed-loop system is stable if the control parameters are chosen to satisfy

$$\left[a + bk_p\left(1 + kh\right)\right]\cdot\left[k_i\left(1 + kh\right) + kk_p\right] + bk^2 k_d k_p - kk_i > 0 \qquad (5\text{-}10)$$

and N is sufficiently large [94]. Furthermore v_r and δ converge exponentially fast to the residual set

$$E = \left\{ v_r \in R, \delta \in R \middle| |v_r| \le C_1 \cdot \left\|\dot{v}_l(t)\right\|_\infty + C_2 \cdot \left\|\dot{d}(t)\right\|_\infty \right.$$
$$\left. \text{and} \qquad |\delta| \le C_3 \cdot \left\|\dot{v}_l(t)\right\|_\infty + C_4 \cdot \left\|\dot{d}(t)\right\|_\infty \right\} \qquad (5\text{-}11)$$

for some finite constants $C_i > 0$, $i = 1,2,3,4$. It is implied by (5-11) that v_r and δ converge to zero exponentially fast if v_l and d are constants.

Using (5-9) the fuel command is issued when u is positive, while the brake is activated when $u < -u_0$ (u_0 is a positive design parameter). Otherwise, the brake system is inactive and the fuel system is operating as in idle speed. The control effort u is multiplied by a fixed gain whenever u is negative in order to deal with the different actuator limits.

Remark 5-2: The simple switching logic described above guarantees that the fuel and brake systems will not be active at the same time and prevents frequent chattering between the two subsystems. The constant u_0 is a design parameter to be chosen for vehicle following performance. The simulation results demonstrate that as long as the control parameter u_0 is properly chosen, the proposed switching logic works in a good manner. It should be noted that the switching logic used here is based on the control effort u rather than the desired acceleration. An alternative switching logic described in [93] is used to deal with the same problem.

The above analysis is for the case that h and k are both constants. In [12], the desired inter-vehicle spacing is taken as

$$s_d = s_0 + h_1 v + h_2 v^2 \tag{5-12}$$

where h_1 and h_2 are positive constants. The time headway used in this desired spacing is given as $h_1+h_2 v$. We refer to (5-12) as the quadratic spacing policy. In [91], the time headway h and the control parameter k are chosen as

$$\begin{cases} h = \text{sat}(h_0 - c_h v_r) \\ k = c_k + (k_0 - c_k)\exp(-\sigma\delta^2) \end{cases} \tag{5-13}$$

where h_0, c_h, k_0, c_k and σ are positive constants to be designed (with $c_k < k_0$) and the saturation function sat($\cdot$) has an upper bound 1 and a lower bound 0. Other choices of spacing rules based on traffic flow characteristics can be found in [77]. If the lead vehicle operates around a nominal speed v_{l0} and the vehicle following controller can efficiently keep v_r and δ close to zero, the stability analysis for the constant time headway policy can be applied to the quadratic spacing policy and the nonlinear spacing policy in (5-13) for small perturbations around the nominal speed v_{l0}.

121

As we mentioned before, based on the simplified model represented by (5-7) and (5-8), different design control methodologies, including the one presented in Chapter 4, can be developed. Our simulation results indicate that for a given heavy truck, its transient speed response characteristics depend on the spacing rule used, i.e. how the time headway h in (5-5) and the control parameter k in (5-9) are chosen. In the simulations, we use the PID form controller in (5-9) for different spacing policies.

To guarantee that the constraints **C1**, **C2** are not violated, we should avoid the generation of high or fast varying control signals. Such high or fast varying control signals can be generated by the control law (5-9) if the lead vehicle accelerates rapidly or changes lanes creating a large spacing error, or the truck switches to a new target with large initial spacing error. In [91, 92], these practical situations have not been addressed since the emphasis was on truck platoons where trucks with very similar dynamic characteristics are assumed and no manually driven vehicles were allowed to interact with the automated trucks. In [28], a nonlinear filter shown in Figure 5-1 is used to smooth the speed trajectory of the lead vehicle, and sat(δ) is used instead of δ. These modifications work well for the passenger vehicle case. However, in the heavy truck case, fast varying v_l may also lead to fast varying δ since the heavy truck can only accelerate slowly, which indicates that δ should also be smoothed before being passed into the control system. Furthermore, in this situation, the temporary separation error could be very large. If we simply use sat(δ) for the controller (5-9), any changes in v_l will directly affect the control signal when $\delta > $ sat(δ). In the following section we design a new vehicle following controller to address these issues in addition to others.

122

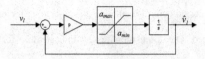

Figure 5-1: Nonlinear filter used to smooth the speed of the lead vehicle.

5.3.4 New Vehicle Following Controller with Improved Performance

Most choices of desired inter-vehicle spacing [12, 66, 77, 83, 91], share the same property

$$H \triangleq \frac{\partial}{\partial v} s_d (v, v_l) \geq 0 \qquad (5\text{-}14)$$

For easy reference, we define

$$H_l \triangleq \frac{\partial}{\partial v_l} s_d (v, v_l) \qquad (5\text{-}15)$$

For the constant time headway rule, H is equal to the time headway h and H_l is zero. In the following analysis, we always assume that H and H_l are bounded since in practice it is impossible to employ time headway involving arbitrarily large H and H_l.

Lemma 5-1: For the vehicle following problem described in section 5.3.1, if the controller is designed such that $(v_r + k\delta) \to 0$ as $t \to \infty$ (k is a positive constant) and $\frac{d}{dt}(v_r + k\delta)$ is uniformly continuous, then v_r and δ are bounded. In addition, if v_l is a constant, then the control objective in (5-6) is achieved.

123

Proof: If $\frac{d}{dt}(v_r+k\delta)$ is uniformly continuous and $(v_r+k\delta)\to0$ as $t\to\infty$, then it follows from Barbalat's Lemma [27] that $\frac{d}{dt}(v_r+k\delta)\to0$ as $t\to\infty$ and therefore

$$\frac{d}{dt}(v_r+k\delta)=(1+kH)\dot{v}_r+kv_r-k(H+H_l)\dot{v}_l\to0 \qquad (5\text{-}16)$$

as $t\to\infty$. It also follows from $\frac{d}{dt}(v_r+k\delta)$ being uniformly continuous and $\frac{d}{dt}(v_r+k\delta)\to0$ as $t\to\infty$ that $\frac{d}{dt}(v_r+k\delta)$ is bounded. Since k is a positive constant, $H>0$, and H, H_l and $\dot{v}_l$ are bounded, it follows that v_r is bounded. In addition, $(v_r+k\delta)$ is uniformly continuous since $\frac{d}{dt}(v_r+k\delta)$ is bounded. Hence it can be shown that $(v_r+k\delta)$ is bounded since it is uniformly continuous and converges to zero. Hence δ is bounded. In particular, if v_l is a constant, then (5-16) implies $v_r\to0$ and therefore $\delta\to0$ which in turn implies that the control objective in (5-6) is achieved.

$\square$

Lemma 5-1 indicates that the vehicle following task can be viewed as a special speed tracking task, in which the desired speed v_d is equal to $v_r+k\delta$. If we can design the vehicle following controller such that $v\to v_r+k\delta$ in a proper way (for example, keeping $\frac{d}{dt}(v_r+k\delta)\to0$ at the same time), then v_r and δ are guaranteed to be bounded, and the control objective in (5-6) is achieved when v_l is a constant.

We propose the speed tracking controller

$$u=f_u^{-1}(v_d)+k_1e_v+k_2+k_3\dot{v}_d \qquad (5\text{-}17)$$

where $e_v=v_d-v$, and the control parameters k_i ($i=1,2,3$) are updated as

124

$$\begin{cases} \dot{k}_1 = \mathrm{Proj}\{\gamma_1 e_v^2\} \\ \dot{k}_2 = \mathrm{Proj}\{\gamma_2 e_v\} \\ \dot{k}_3 = \mathrm{Proj}\{\gamma_3 e_v \dot{v}_d\} \end{cases} \tag{5-18}$$

with initial values $k_i(0)=k_{i0}$ (i=1,2,3). In (5-18) γ_i (i=1,2,3) are positive design parameters, and Proj$\{\cdot\}$ is the projection function limiting k_i between their lower bounds k_{li} and upper bounds k_{ui} (i =1,2,3).

Lemma 5-2: Consider the system represented in (5-7) and (5-8) with the adaptive speed tracking controller described by (5-17) and (5-18). If k_{li} and k_{ui} (i =1,2,3) are properly chosen, then all the signals inside the closed-loop system are bounded. In addition:

(i) If d is a constant, $e_v \rightarrow 0$ as $t \rightarrow \infty$.

(ii) If d is a constant and $\dot{v}_d$ is uniformly continuous, $e_v, \dot{e}_v \rightarrow 0$ as $t \rightarrow \infty$.

Proof: For the system represented in (5-7) and (5-8), if a, b and d are known, then the controller

$$u = f_u^{-1}(v_d) + k_1^* e_v + k_2^* + k_3^* \dot{v}_d \tag{5-19}$$

where $k_1^* = \dfrac{a_m - a}{b}$, $k_2^* = -\dfrac{d}{b}$, $k_3^* = \dfrac{1}{b}$ and a_m is a positive constant could be used. In this case the close-loop system becomes

$$\dot{e}_v = -a_m e_v \tag{5-20}$$

Hence, $e_v, \dot{e}_v \rightarrow 0$ as $t \rightarrow \infty$. Since a, b and d are unknown, the control law (5-18) is proposed, and the closed-loop system can be rewritten as

125

$$\dot{e}_v = -a_m e_v - \tilde{k}_1 e_v - \tilde{k}_2 - \tilde{k}_3 \dot{v}_d \qquad (5\text{-}21)$$

where $\tilde{k}_i = k_i - k_i^*$ (i=1,2,3). Consider the following Lyapunov function

$$V = \frac{e_v^2}{2} + \frac{b\tilde{k}_1^2}{2\gamma_1} + \frac{b\tilde{k}_2^2}{2\gamma_2} + \frac{b\tilde{k}_3^2}{2\gamma_3} \qquad (5\text{-}22)$$

If k_{li} and k_{ui} are chosen so that $k_{li} \leq k_i^* \leq k_{ui}$ and $k_{li} \leq k_{i0} \leq k_{ui}$ for each i, then the update law given by (5-18) guarantees that $k_{li} \leq k_i \leq k_{ui}$ and [27]

$$\dot{V} = e_v \dot{e}_v + \frac{b}{\gamma_1}\tilde{k}_1 \dot{\tilde{k}}_1 + \frac{b}{\gamma_2}\tilde{k}_2 \dot{\tilde{k}}_2 + \frac{b}{\gamma_3}\tilde{k}_3 \dot{\tilde{k}}_3 \leq -a_m e_v^2 + \frac{1}{\gamma_2}\tilde{k}_2 \dot{d} \qquad (5\text{-}23)$$

It can be shown [27] that all the signals inside the closed-loop are bounded. (5-23) also implies that if d is a constant, then $e_v \in L_2 \cap L_\infty$ and $\dot{e}_v \in L_\infty$, which means $e_v \rightarrow 0$. Furthermore, if $\dot{v}_d$ is uniformly continuous, it can be verified that $\dot{e}_v$ is also uniformly continuous. Using Barbalat's Lemma, we can show that $\dot{e}_v$ also converges to zero.

☐

Since the desired speed $v_l + k\delta$ may vary fast, we employ the nonlinear filter in Figure 5-2 to generate a smooth signal v_{ref} to be tracked. The saturation function inside the nonlinear filter serves as an acceleration limiter that restricts the change rate of v_{ref} to be between a_{min} and a_{max}. The signal generated by the acceleration limiter is z=sat$\{p(v_l + k\delta - v_{ref})\}$, where p is a positive design parameter. The function after the acceleration limiter is designed to accept or ignore the change rate signal z, and is given as

126

$$f\left(z, v_{ref}, v_l\right) = \begin{cases} z, & \text{if } v_l + m_v \leq v_{ref} \leq v_l + M_v \text{ and } z < 0; \\ & \text{or } v_l - m_v < v_{ref} < v_l + m_v; \\ & \text{or } v_{ref} \leq v_l - m_v \text{ and } z > 0 \\ 0, & \text{if } v_l + m_v \leq v_{ref} \leq v_l + M_v \text{ and } z \geq 0; \\ & \text{or } v_{ref} \leq v_l - m_v \text{ and } z \leq 0 \\ a_{\min}, & \text{if } v_{ref} > v_l + M_v \end{cases}$$ (5-24)

where m_v and M_v are constant design parameters with $0 < m_v < M_v$. The purpose of this nonlinear filter is to limit the change rate of v_{ref} to be between $a_{\min}$ and $a_{\max}$, and prevent v_{ref} from becoming much higher or lower than v_l. By regulating the truck's speed towards v_{ref}, the ACC system forces the truck to follow the preceding vehicle in a safe and comfortable manner, while meeting the control objective given by (5-6).

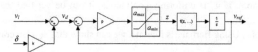

Figure 5-2: Nonlinear filter used in the new vehicle following controller.

The acceleration limits $a_{\min}$ and $a_{\max}$ are chosen based on the truck dynamic capabilities and driver comfort. In our case in order to smooth the truck speed response further these limits are varied based on which region v_r and δ are located in the δ-v_r plane. As shown in Figure 5-3, all the trajectories in the δ-v_r plane should be regulated towards the dashed line described by $v_r + k\delta = 0$, and finally reach the origin ($v_r = 0$, $\delta = 0$) if the speed of the lead vehicle is a constant. The solid line in the δ-v_r plane representing $v_r + k\delta = -B$ ($B > 0$) splits the plane into two regions: the safe region (region 1) and the unsafe region

(region 2). It is expected that the ACC truck should work in region 1 most of the time. In this region, $a_{\min}$ and $a_{\max}$ are chosen so that the constraints **C1** and **C2** are satisfied and the truck actuation systems will not be saturated when the truck speed tightly follows v_{ref}. In region 2, $v_r + k\delta \leq -B$ indicates that the desired speed $v_d = v_r + k\delta$ is much smaller than v. In this region the absolute values used for $a_{\min}$ are larger than those used in region 1. In other words, v_{ref} is allowed to decrease faster in order to guarantee safety provided of course the acceleration limits based on driver comfort and characteristics of the truck are not violated.

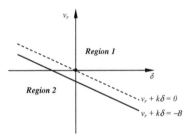

Figure 5-3: Region 1 and Region 2 in the δ-v_r plane.

Remark 5-3: Even though the signal z within the nonlinear filter in Figure 5-2 is continuous, the function (5-24) may generate discontinuous signals that may cause problems in the analysis related to the existence and uniqueness of solutions of the resulting differential equation. The discontinuities may arise when v_{ref} varies around $v_l - m_v$, or $v_l + m_v$ or $v_l + M_v$. However, the function (5-24) can be slightly modified so that it will always generate continuous signals when z is continuous. For example, we can choose a small positive constant ε, and when $z>0$ and $v_l+m_v-\varepsilon \leq v_{ref} < v_l+m_v$ are satisfied, we set f equal to $(v_l+m_v-v_{ref}) \cdot z/\varepsilon$ instead of z. This minor modification will guarantee that no discontinuous signals are generated when v_{ref} varies around v_l+m_v.

Lemma 5-3: Consider the system (5-7) and (5-8) with the controller given by (5-17) and (5-18) with v_d replaced with v_{ref}. If k_{li} and k_{ui} ($i = 1,2,3$) are properly chosen and v_{ref} is generated by the nonlinear filter in Figure 5-2, then u, v and v_r are bounded. In addition:

(i) If d is a constant, $v \to v_{ref}$ and $\dot{v} \to \dot{v}_{ref}$ as $t \to \infty$.

(ii) If v_l and d are constants, and the control parameters are chosen such that

$$(1/k + \inf H)|a_{\min}| > m_v \qquad (5\text{-}25a)$$

$$(1/k + \inf H)a_{\max} > m_v \qquad (5\text{-}25b)$$

where $\inf H$ is the infimum of H, then all the signals are bounded.

Proof: By considering the function in (5-24), it is easy to see that $|v_l-v_{ref}|$ is bounded from above by $\max\{|v_l(0)-v_{ref}(0)|, M_v\}$. Since we can always set $v_l(0)-v_{ref}(0)=0$, we conclude that $|v_l-v_{ref}|$ is bounded by M_v. It follows from Lemma 5-2 that $(v-v_{ref})$ is bounded which implies that v_r is bounded. Using the fact that v_l is bounded, it is easy to show that u, v are bounded.

It can be shown that $\frac{d}{dt}(v_l+k\delta)$ is bounded, so it follows that $v_l+k\delta$ is uniformly continuous. It is verified that $\dot{v}_{ref}$ generated by (5-24) (with the modifications suggested in Remark 5-3) is uniformly continuous. Hence part (i) can be proven using Lemma 5-2.

For part (ii), when v_l and d are constants, we consider the Lyapunov function:

129

$$V = \tfrac{1}{2} x^T P x \qquad (5\text{-}26)$$

where $P = \begin{bmatrix} 1/k + 1/p & 1 \\ 1 & k \end{bmatrix} > 0$, and $x = [v_I - v_{ref}, \; \delta]^T$. We have

$$\dot{V} = -\left[(1/k + 1/p + H) x_1 + (1 + kH) x_2 \right] \dot{x}_1 \\ + (x_1 + kx_2) x_1 + (x_1 + kx_2)(\eta_1 + H\eta_2) \qquad (5\text{-}27)$$

where $\eta_1 = v_{ref} - v$ and $\eta_2 = \dot{v}_{ref} - \dot{v}$. In the following analysis, we only consider the situations where $v_I - m_v \leq v_{ref} \leq v_I + m_v$ is satisfied. This is valid since by design v_{ref} will be bounded by $v_I - m_v$ and $v_I + m_v$ in finite time for any bounded initial condition.

① $v_I - m_v < v_{ref} < v_I + m_v$, or $v_{ref} = v_I + m_v$ and $z \leq 0$, or $v_{ref} = v_I - m_v$ and $z \geq 0$. In this case, $\dot{x}_1 = -z = -\mathrm{sat}\{p(x_1 + kx_2)\}$. Hence

$$\dot{V} = -\left[(1/k + 1/p + H) x_1 + (1 + kH) x_2 \right] \mathrm{sat}\{p(x_1 + kx_2)\} \\ + (x_1 + kx_2) x_1 + (x_1 + kx_2)(\eta_1 + H\eta_2) \\ = -(1/k + H)(x_1 + kx_2)\mathrm{sat}\{p(x_1 + kx_2)\} \\ - (1/p) x_1 \mathrm{sat}\{p(x_1 + kx_2)\} + (x_1 + kx_2) x_1 + (x_1 + kx_2)(\eta_1 + H\eta_2) \qquad (5\text{-}28)$$

If $a_{min} \leq p(x_1 + kx_2) \leq a_{max}$, then (5-28) becomes

$$\dot{V} = -p(1/k + H)(x_1 + kx_2)^2 + (x_1 + kx_2)(\eta_1 + H\eta_2) \qquad (5\text{-}29)$$

Hence $\dot{V} < 0$ if $|x_1 + kx_2| > |\eta_1 + \eta_2 H|/[p(1/k+H)]$. If $p(x_1 + kx_2) > a_{max}$, then (5-28) becomes

$$\dot{V} = -(1/k + H)(x_1 + kx_2) a_{max} - (1/p) x_1 a_{max} \\ + (x_1 + kx_2) x_1 + (x_1 + kx_2)(\eta_1 + H\eta_2) \qquad (5\text{-}30)$$

It is easy to verify that when t is sufficiently large and (5-25b) is satisfied, $\dot{V} < 0$ always holds. If $p(x_1+kx_2) < a_{\min}$, it can also be verified that when t is sufficiently large and (5-25a) is satisfied, $\dot{V} < 0$ always holds.

② $v_{ref}=v_l+m_v$ and $z>0$. In this case, $x_1=-m_v$, $x_1+kx_2\geq z>0$ and $\dot{x}_1=-z=0$. Hence,

$$\dot{V} = -(x_1 + kx_2)m_v + (x_1 + kx_2)(\eta_1 + H\eta_2) \tag{5-31}$$

Hence when t is sufficiently large, $m_v > |\eta_1 + \eta_2 H|$ always holds, which indicates that $\dot{V} < 0$ always holds.

③ $v_{ref} = v_l + m_v$ and $z>0$. Following the same way as in ②, it can be shown that $\dot{V} < 0$ always holds.

As $t\to\infty$, $\dot{V}$ is always non-positive when $|x_1+kx_2| > |\eta_1+\eta_2 H|/[p(1/k+H)]$. Since η_1, η_2, H and x_1 are bounded, it can be established that there exists a positive number $V_{\max}$ such that $\dot{V} < 0$ if $V > V_{\max}$. Hence we can conclude that V is bounded and then all the signals inside the closed-loop system are bounded.

□

Remark 5-4: In the proof of Lemma 5-3, we have assumed that (5-24) always generates continuous signals when z is continuous. One can verify that using the modifications for (5-24) suggested in Remark 5-3, the proof for Lemma 5-3 can be achieved in a similar way but with more regions for v_{ref}.

Remark 5-5: If η_1 and η_2 in (5-27) are zero, it can be shown that x_1, $x_2 \rightarrow 0$ as $t \rightarrow \infty$, i.e. the control objective in (5-6) is achieved. The simulation results demonstrate that (5-6) is achieved when v_l is a constant.

In summary, the new ACC system is formed by the reference speed generator in Figure 5-2 and the speed tracking controller given in (5-17) and (5-18) with v_d replaced with v_{ref}. By tracking the smooth speed trajectory generated by the nonlinear filter in Figure 5-2, the ACC truck will not attempt to track speed responses of the lead vehicle that involve high frequency oscillations and/or high accelerations. As demonstrated by the simulation results in section 5.4, the new ACC system can effectively prevent the actuator saturation problem for heavy trucks and attenuate certain speed disturbances in mixed traffic situations. In [6], Bae and Gerdes have used a different technique called "input shaping" to prevent the actuator saturation problem and improve vehicle following performance for truck platooning.

In addition to the nonlinear shaping function of the reference speed, the following switching rules are incorporated in the new ACC system:

S1. If the separation distance x_r is larger than x_{max} (x_{max} is a positive design constant), then the fuel system is on.

S2. If the separation distance x_r is smaller than x_{min} (x_{min} is a positive design constant), then the brake system is on.

S3. $x_{min} \leq x_r \leq x_{max}$, then the fuel system is on when $u > 0$, while the brake is activated when $u < -u_0$. When $-u_0 \leq u \leq 0$, the brake system is inactive and the fuel system is operating as in idle speed.

S1 and *S2* in the vehicle following mode are used to avoid unnecessary activities of the brake or fuel systems when the vehicle separation is large or small enough. They are similar to those used in [28].

5.4 Simulation Studies

Simulations are carried out using Matlab/Simulink to investigate how different trucks would affect the speed and separation responses of the following passenger vehicles. The dynamics of the ACC trucks are modeled using the validated nonlinear model proposed in [89, 90], the human drivers for passenger cars are modeled using the Pipes' model and the drivers for trucks are modeled using the modified Bando's model.

For easy reference, we use ACC_N, ACC_Q, ACC_C and ACC_NEW to represent automated trucks with the nonlinear spacing rule in (5-13), quadratic spacing rule, constant time headway rule and the newly developed controller, respectively.

We simulate and compare six vehicle strings, each containing ten vehicles. The lead vehicle in the six strings generates the same speed trajectory to be followed by the vehicles in the string. The second vehicle in the six strings is a manually driven passenger vehicle (modeled using the Pipes' model), manually driven truck (modeled using the modified Bando's model), ACC_N, ACC_Q, ACC_C and ACC_NEW, respectively. The

133

other eight vehicles in each string are manually driven passenger vehicles following the second vehicle. The parameters for the ACC systems are chosen as:

ACC_N: $h_0=1.6$, $k_0=1$, $c_h=0.2$, $c_k=0.5$, $\sigma=0.1$, $s_0=6.0$m

ACC_Q: $h_1=0.8$, $h_2=0.03$, $k=0.2$ and $s_0=6.0$m

ACC_C: $h=1.6$, $k=0.2$ and $s_0=6.0$m

ACC_NEW: $h=1.6$, $k=0.2$, $s_0=6.0$m, $p=10$, $m_v=2.0$m/sec, $M_v=8.0$m/sec, and a_{min} and a_{max} used for the nonlinear filter in Figure 5-2 are acceleration limits chosen based on the capabilities of the truck and desired driver comfort considerations.

In the simulations, the truck weight is fixed at 20 tons. In ACC_N, ACC_Q and ACC_C, the speed of the lead vehicle is processed with the nonlinear filter shown in Figure 5-1, and the acceleration bounds in the nonlinear filter are chosen to be the same as those used in the nonlinear filter shown in Figure 5-2. We also use sat(δ) in ACC_Q and ACC_C to eliminate the adverse effect of large separation errors. The PID parameters for the ACC controllers are tuned for each truck to achieve good vehicle following performance. It should be noted that we cannot limit the rate of change of the control effort by simply decreasing the PID gains for ACC_N, ACC_Q and ACC_C since small control gains will deteriorate the vehicle following performance due to the slinky effect. A low-pass filter is placed after each ACC controller to limit the change rate of control effort.

In our simulations, we assume that the ranging sensor installed on ACC trucks to measure inter-vehicle distance and relative speed has a maximum operation range of 120 meters. If the vehicle separation distance is larger than 120 meters, the ACC system will switch from the vehicle following mode to the speed tracking (self-cruising) mode.

134

5.4.1 Low Acceleration Maneuvers

In these simulations, the lead passenger vehicle accelerates from 8m/sec to 20m/sec with a constant acceleration of $0.8m/sec^2$, and then cruises at a constant speed. The $0.8m/sec^2$ acceleration is easily achievable by the passenger vehicles but it is larger than the maximum acceleration a heavy truck could achieve with the particular trailer mass used in the simulations. When the lead passenger vehicle increases its speed with accelerations much smaller than $0.8m/sec^2$, a heavy truck can easily follow the preceding vehicle. The five different trucks will therefore behave similar to a passenger vehicle. However, if the lead passenger vehicle increases its speed faster than $0.8m/sec^2$, the responses of the manually driven truck and the four ACC trucks are much different.

Figures 5-4(a) and 5-4(b) show the speed and separation error responses in vehicle string 1 (all manually driven passenger vehicles), and Figures 5-5(a) and 5-5(b) show the speed and separation error responses in vehicle string 2 (manually driven passenger vehicles with the second vehicle in the string being a manually driven truck). The response labeled by "vi" corresponds to the ith vehicle in the string. These results indicate that the manually driven truck behavior is similar to that of the second passenger vehicle in vehicle string 1. The truck however has to reach its maximum acceleration limit in order to follow the preceding vehicle, leading to a slightly larger overshot in the speed response. The maximum transient separation error of the truck is slightly larger than that of the second passenger vehicle in vehicle string 1.

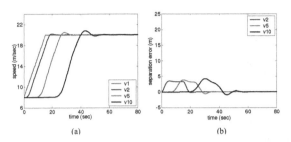

(a)   (b)

Figure 5-4: Low acceleration: (a) speed responses and (b) separation error responses of the vehicles in string 1 (all the vehicles are manually driven passenger vehicles).

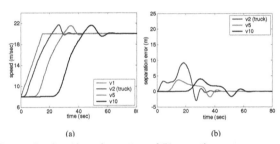

(a)   (b)

Figure 5-5: Low acceleration: (a) speed responses and (b) separation error responses of the vehicles in string 2 (the second vehicle is a manually driven heavy truck).

Figures 5-6, 5-7 and 5-8 show the speed, separation error and truck fuel responses in vehicle strings 3, 5 and 6, respectively. The responses in vehicle string 4 are not presented here since ACC_Q behaves very similarly to ACC_C, and they are not presented in the following two sub-sections, either. The vehicle following controllers in ACC_N, ACC_Q and ACC_C tend to generate large control efforts in the presence of large relative speeds and separation errors, which easily leads to fuel system saturation. Within the three controllers, the one in ACC_N has the most aggressive behavior, due to

136

its variable time headway. However, the newly developed ACC controller regulates the truck's speed towards a smooth reference speed generated by the nonlinear filter in Figure 5-2. Therefore, its fuel system responds in a smooth way. The fuel saturation is avoided in Figure 5-8(c) and the truck's acceleration is kept within the desired acceleration limits a_{min} and a_{max}. The adverse effect caused by the new ACC controller is that it makes the heavy truck response a little more sluggish, and the transient separation error is larger than those of the other ACC trucks until the truck catches up with the lead vehicle.

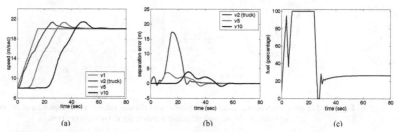

(a)   (b)   (c)

Figure 5-6: Low acceleration: (a) speed responses and (b) separation error responses of the vehicles in string 3 (the second vehicle is an ACC truck with the nonlinear spacing rule), and (c) fuel response the ACC truck.

137

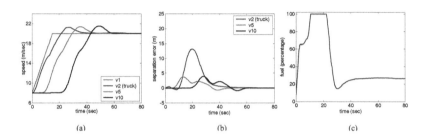

(a)   (b)   (c)

Figure 5-7: Low acceleration: (a) speed responses and (b) separation error responses of the vehicles in string 5 (the second vehicle is an ACC truck with the constant time headway spacing rule), and (c) fuel response the ACC truck (The responses in string 4 are very similar).

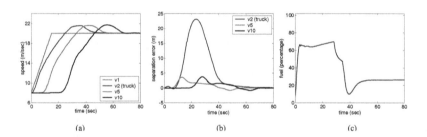

(a)   (b)   (c)

Figure 5-8: Low acceleration: (a) speed responses and (b) separation error responses of the vehicles in string 6 (the second vehicle is an ACC truck with the new controller), and (c) fuel response the ACC truck.

138

5.4.2 High Acceleration Maneuvers

In this case, we want to investigate how heavy trucks would affect the responses of the following passenger vehicles when the lead vehicle accelerates rapidly. In the simulations, the lead passenger vehicle accelerates from 8m/sec to 20m/sec with a constant acceleration of 2m/sec^2, and then cruises at a constant speed.

Figures 5-9 and 5-10 show the speed and separation error responses in vehicle strings 1 and 2, respectively. The separation error for the manual truck is temporarily large due to its inability to accelerate as fast as the lead passenger vehicle. The truck driver regulates the vehicle speed to be higher than that of the lead vehicle in order to close in and maintain the desired separation distance. This leads to a peak in the speed response as shown in Figure 5-10(a). It indicates that the slinky effect of a manually driven heavy truck is more serious than that of a passenger vehicle. Figures 5-11, 5-12 and 5-13 show the speed, separation error and truck fuel responses in vehicle strings 3, 5 and 6, respectively. Figures 5-11(c) and 5-12(c) indicate that the traditional ACC controllers lead to fuel system saturation very soon in order to track the speed of the lead vehicle and maintain the desired separation. The new ACC controller regulates the truck's speed in a smooth way without having to saturate the fuel system, and the control objective in (5-6) can be achieved eventually, as shown in Figure 5-13.

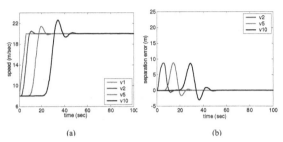

(a) (b)

Figure 5-9: High acceleration: (a) speed responses and (b) separation error responses of the vehicles in string 1 (all the vehicles are manually driven passenger vehicles).

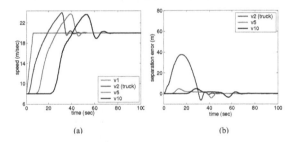

(a) (b)

Figure 5-10: High acceleration: (a) speed responses and (b) separation error responses of the vehicles in string 2 (the second vehicle is a manually driven heavy truck).

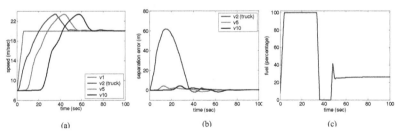

(a) (b) (c)

Figure 5-11: High acceleration: (a) speed responses and (b) separation error responses of the vehicles in string 3 (the second vehicle is an ACC truck with the nonlinear spacing rule), and (c) fuel response the ACC truck.

140

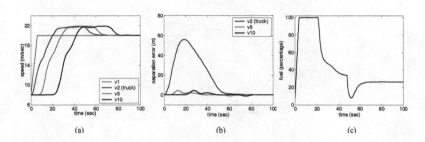

(a)   (b)   (c)

Figure 5-12: High acceleration: (a) speed responses and (b) separation error responses of the vehicles in string 5 (the second vehicle is an ACC truck with the constant spacing rule), and (c) fuel response the ACC truck (The responses in string 4 are very similar).

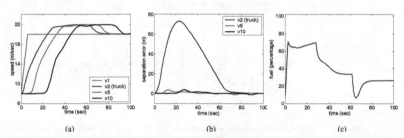

(a)   (b)   (c)

Figure 5-13: High acceleration: (a) speed responses and (b) separation error responses of the vehicles in string 6 (the second vehicle is an ACC truck with the new controller), and (c) fuel response the ACC truck.

141

5.4.3 High Acceleration Maneuvers with Oscillations

In this case we simulate the situation where the lead passenger vehicle accelerates from 8m/sec to 16m/sec with a constant acceleration of 2.0m/sec^2, and its speed oscillates around 16m/sec before settling to the constant speed of 16m/sec. This situation may arise in today's traffic where traffic disturbances downstream create a situation where the driver speeds up and then slows down in an oscillatory fashion before reaching steady state. The speed oscillations may also arise due to lane change behaviors before the lead vehicle. In this situation, the temporary separation distance between the first and second vehicles will be very large, and we would like to investigate how the vehicles in the various strings of vehicles considered will behave with respect to the speed oscillations of the lead vehicle.

Figures 5-14, 5-15, 5-16, 5-17 and 5-18 show the responses in vehicle strings 1, 2, 3 and 5, respectively. In Figures 5-14 and 5-15, the speed disturbance introduced by the lead vehicle is transferred upstream unattenuated due to the aggressive behavior of human drivers. Figure 5-16 shows that the responses of ACC_N are similar to those of the manually driven truck. ACC_C behaves better than ACC_N. However, when its speed is larger than that the lead vehicle, the speed disturbance is transferred upstream unattenuated. This is due to the use of sat(δ). The control effort generated by (5-9) will be directly affected by v_l, since sat(δ) is always a constant when δ is large. The same is true for ACC_Q. Figure 5-19 shows the speed, separation error and truck fuel responses in vehicle string 6. In this case due to the new ACC truck controller, the speed disturbance is attenuated. This attenuation is due to the use of (5-24). Note, the speed disturbance can only be attenuated when the separation error is positive. When the separation error is negative, the ACC truck's response will be to reduce the separation error for safety reasons rather than reduce the oscillations.

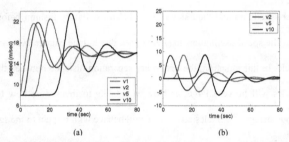

(a)   (b)

Figure 5-14: High acceleration with oscillations: (a) speed responses and (b) separation error responses of the vehicles in string 1 (all the vehicles are manually driven passenger vehicles).

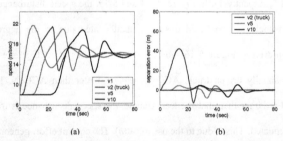

(a)   (b)

Figure 5-15: High acceleration with oscillations: (a) speed responses and (b) separation error responses of the vehicles in string 2 (the second vehicle is a manually driven heavy truck).

143

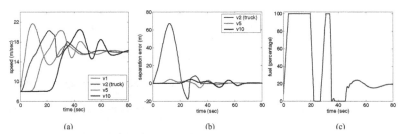

Figure 5-16: High acceleration with oscillations: (a) speed responses and (b) separation error responses of the vehicles in string 3 (the second vehicle is an ACC truck with the nonlinear spacing rule).

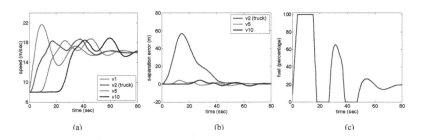

Figure 5-17: High acceleration with oscillations: (a) speed responses and (b) separation error responses of the vehicles in string 5 (the second vehicle is an ACC truck with the constant time headway spacing rule), and (c) fuel response the ACC truck (The responses in string 4 are very similar).

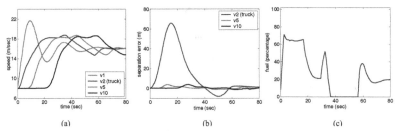

Figure 5-18: High acceleration with oscillations: (a) speed responses and (b) separation error responses of the vehicles in string 6 (the second vehicle is an ACC truck with the new controller), and (c) fuel response the ACC truck.

5.5 fuel economy and emissions

In this section, we investigate how different heavy trucks, manually driven or ACC operated, affect the fuel economy and emissions of the passenger vehicles in the presence of different traffic disturbances simulated in the previous section. We also investigate how different ACC systems affect the fuel efficiency and emissions of the heavy trucks in the simulated scenarios. In addition, we investigate how the new proposed ACC system denoted by ACC_NEW affects the following vehicles in the presence of speed limits or cut-in vehicles.

In the emission analysis, the quantities measured are the tailpipe emissions of unburned hydrocarbons (HC), monoxide of carbon (CO), oxides of nitrogen (NO, NO_2, denoted by NO_x) and fuel consumption. For passenger vehicles, the Comprehensive Modal Emissions Model (CMEM) developed at UC Riverside is used to analyze the vehicle data and calculate the air pollution and fuel consumption [8]. The vehicle category is set to be 5, which is the most common vehicle type in California: high-mileage, high power-to-weight. The emission prediction model used for heavy trucks is developed by the same research group [67]. The inputs for the two models are vehicle longitudinal speed and acceleration data, while road grade is taken to be zero and no wind gust is considered. The outputs generated by the two emission models include tailpipe emissions of HC, CO, and NO_x, and fuel consumption.

5.5.1 Low Acceleration Maneuvers

The purpose of this test is to examine the effect of the truck (2nd in the string) on the behavior of the following 8 passenger cars when the lead vehicle performs a low acceleration maneuver creating a disturbance that propagates upstream. The lead passenger vehicle accelerates from 8m/sec to 20m/sec with a constant acceleration of $0.8 m/sec^2$, and then cruises at a constant speed. We calculate the fuel consumption and emissions of each vehicle from the time the lead vehicle begins to accelerate, until the string covers a distance of 1.2km, as this is the distance taken by the vehicles to reach a steady state speed after completing the acceleration maneuver. The travel time is also recorded over a distance of 1.2km.

String of 10 vehicles	Manual Passenger Vehicle in the 2nd position	Manual truck in the 2nd position	ACC_N in the 2nd position	ACC_Q in the 2nd position	ACC_C in the 2nd position	ACC_NEW in the 2nd position
Travel Time (second)	73.7	73.9	73.3	73.2	73.3	73.3
Fuel (g)	0% (617.8)	≈ 0	≈ 0	≈ 0	≈ 0	4.3%
CO (g)	0% (20.6)	≈ 0	7.3%	6.4%	6.0%	9.4%
HC (g)	0% (0.93)	≈ 0	7.1%	6.3%	5.9%	9.2%
NO_x (g)	0% (1.51)	9.1%	10.0%	12.1%	10.7%	24.5%

Table 5-1: Travel time, fuel and emission benefits of the 8 passenger vehicles following the truck in a string of 10 vehicles for low acceleration maneuvers of the lead vehicle (no cut-ins).

Table 5-1 shows the travel times and fuel and emission benefits for the last eight passenger vehicles in each vehicle string in the presence of different truck in the 2nd position in the string of the 10 vehicles. The data in the second column are for vehicle string 1 (all manually driven passenger vehicles), and the fuel and emission benefits are denoted by 0% since this case is used as the basis for comparison. The data in columns 3-

7 are for vehicle strings 2-6, respectively. For the fuel and emission data, a positive number means improvement while a negative number means deterioration. The notation "≈ 0" in the table represents benefits between –4% and 4%, and is considered to be negligible due to inaccuracies in the emission model. Due to the aggressive response of the manual truck, the fuel and emission benefits are minor except for NO_x. The smooth response of the ACC trucks filters the disturbance created by the lead vehicle presenting a smoother speed response to be tracked by the passenger vehicles following the truck. As a result some benefits with respect to fuel and emissions are shown in Table 5-1. These benefits are small due to the fact that the disturbance created was a rather smooth one (low acceleration). In this case the travel time is not affected.

We also evaluate the fuel consumption and emission results for heavy trucks in these simulations and the data are presented in Table 5-2. The ACC trucks lead to better fuel consumption and emission results compared to the manually driven truck. However, there is little difference among the four different ACC trucks. The benefits are small however due to low level of disturbance.

	Manual Truck	ACC_N	ACC_Q	ACC_C	ACC_NEW
Fuel (g)	0% (594.8)	5.8%	6.0%	5.4%	6.6%
CO (g)	0% (5.27)	5.3%	5.3%	5.1%	6.2%
HC (g)	0% (0.26)	≈ 0	≈ 0	≈ 0	≈ 0
NO_x (g)	0% (12.42)	4.8%	4.8%	4.2%	5.8%

Table 5-2: Fuel and emission benefits of the heavy truck during low acceleration maneuvers of the lead vehicle (no cut-ins).

5.5.2 High Acceleration Maneuvers

The purpose of this test is to examine the effect of the truck (2nd in the string) on the behavior of the following 8 passenger cars when the lead vehicle performs a high acceleration maneuver creating a disturbance that propagates upstream. The lead passenger vehicle accelerates from 8m/sec to 20m/sec with a constant acceleration of 2.0m/sec^2, and then cruises at a constant speed. As in test 1, we consider the same six strings of vehicles with 10 vehicles in each string. The travel time is recorded over a distance of 1.7km as this is the distance taken by the vehicles to reach a steady state speed after completing the acceleration maneuver.

String of 10 vehicles	Manual Passenger Vehicle in the 2nd position	Manual truck in the 2nd position	ACC_N in the 2nd position	ACC_Q in the 2nd position	ACC_C in the 2nd position	ACC_NEW in the 2nd position
Travel Time (second)	96.0	96.2	95.4	95.3	95.4	95.4
Fuel (g)	0% (1020)	15.5%	21.5%	22.7%	22.4%	23.8%
CO (g)	0% (251.2)	87.9%	89.4%	89.9%	89.9%	90.2%
HC (g)	0% (4.02)	65.1%	68.4%	71.5%	71.4%	73.1%
NO$_x$ (g)	0% (4.21)	52.1%	56.0%	60.6%	60.3%	67.6%

Table 5-3: Travel time, fuel and emission benefits of the 8 passenger vehicles following the truck in a string of 10 vehicles for high acceleration maneuvers of the lead vehicle (no cut-ins).

Table 5-3 shows the travel time and fuel consumption and emission results for the last eight passenger vehicles in the simulations with high acceleration maneuvers. The presence of heavy trucks little affects the travel times, but significantly increases the fuel efficiency and decreases emissions of the following passenger vehicles. The trucks act as low-pass filters presenting a much smoother speed response to be tracked by the vehicles following the trucks in the string. Furthermore, it is observed that all the ACC trucks lead

148

to better fuel consumption and emissions compared with the manual truck. The new ACC system has the best improvements.

Table 5-4 shows the fuel consumption and emission results for different trucks. All the ACC trucks have better results in terms of fuel efficiency and emissions compared with the manually driven truck. This is due to the smoother response of the ACC trucks compared with the more aggressive manually driven truck. However, there is little difference among the four different ACC trucks.

	Manual Truck	ACC_N	ACC_Q	ACC_C	ACC_NEW
Fuel (g)	0% (802.0)	10.6%	12.7%	12.6%	12.6%
CO (g)	0% (7.09)	9.2%	11.1%	11.0%	11.2%
HC (g)	0% (0.34)	4.4%	5.2%	5.1%	5.2%
NO$_x$ (g)	0% (18.14)	7.6%	9.5%	9.5%	10.5%

Table 5-4: Fuel and emission benefits of the heavy trucks in high acceleration maneuvers (no cut-ins).

5.5.3 High Acceleration Maneuvers with Oscillations

The purpose of this test is to examine the effect of the truck (2nd in the string) on the behavior of the following 8 passenger cars when the lead vehicle performs a high acceleration oscillatory maneuver creating a disturbance that propagates upstream. The lead passenger vehicle accelerates from 8m/sec to 16m/sec with a constant acceleration of 2.0m/sec^2, and its speed oscillates around 16m/sec before settling to the constant speed of 16m/sec. The travel time is recorded over a distance of 1.3km as this is the distance taken by the vehicles to reach a steady state speed after completing the acceleration maneuver.

Table 5-5 shows the fuel consumption and emission results for the last eight passenger vehicles in each vehicle string. It follows from Table 5-5, that the presence of heavy trucks can significantly improve the fuel efficiency and decrease most of the emissions. This is due to the fact that the inherent sluggish characteristics of the heavy trucks can attenuate the high frequency components in the speed of the lead vehicle. Among the ACC trucks, the ACC_NEW leads to the best results in all aspects, since it filters the oscillations in the lead vehicle's speed more effectively than the others. Again we can observe the travel times are little affected by the heavy trucks.

String of 10 vehicles	Manual Passenger Vehicle in the 2nd position	Manual truck in the 2nd position	ACC_N in the 2nd position	ACC_Q in the 2nd position	ACC_C in the 2nd position	ACC_NEW in the 2nd position
Travel Time (second)	88.0	88.1	87.5	87.5	87.6	87.6
Fuel (g)	0% (860.3)	23.9%	29.7%	34.2%	34.0%	36.1%
CO (g)	0% (298.3)	90.2%	93.5%	94.5%	94.5%	94.9%
HC (g)	0% (4.45)	73.7%	79.3%	82.4%	82.8%	84.1%
NO_x (g)	0% (4.87)	59.4%	65.0%	68.8	69.0%	78.6%

Table 5-5: Travel time, fuel and emission benefits of the 8 passenger vehicles following the truck in a string of 10 vehicles for high acceleration maneuvers with oscillations of the lead vehicle (no cut-ins).

	Manual Truck	ACC_N	ACC_Q	ACC_C	ACC_NEW
Fuel (g)	0% (618.2)	12.0%	23.1%	22.9%	27.4%
CO (g)	0% (5.55)	10.6%	20.0%	20.0%	23.8%
HC (g)	0% (0.28)	4.8%	8.5%	8.5%	12.4%
NO_x (g)	0% (13.2)	9.1%	17.0%	17.2%	21.6%

Table 5-6: Fuel and emission benefits of the trucks in high acceleration maneuvers with oscillations (no cut-ins).

Table 5-6 shows the fuel consumption and emissions results for different trucks. All the ACC trucks lead to better fuel consumption and emissions results compared with the manually driven truck, and the performance of the new ACC system is the best.

5.5.4 Speed Limit Effect

In the previous simulations, we assume that the lead vehicle reaches a steady state giving time for the truck to close in by using a higher speed leading to a travel time that is not affected by the presence of the truck. However, if the lead vehicle reaches a speed that the truck cannot exceed due to load or speed limits, the travel time will be affected. In the simulations, we consider two vehicle strings: string 1 (all manually driven passenger vehicles) and string 6 (with ACC_NEW in the second position) used in the previous simulations. The lead passenger vehicle accelerates from 8m/sec to 24m/sec with a constant acceleration of $1.0m/sec^2$, and then cruises at a constant speed. Here we assume the speed limit is 24m/sec. In string 1, all the following passenger vehicles respond similarly to those shown in Figure 5-4 and the data are not presented here. The speed and separation error responses in string 6 are shown in Figure 5-19. The ACC_NEW increases its speed sluggishly and the separation error keeps increasing until its speed reaches the limit. As marked in Figure 5-19, at time T_{swich}, the inter-vehicle distance becomes larger than the maximum operating range of the ranging sensor and the ACC controller switches from the vehicle following mode to the speed tracking mode. The separation error response is plotted with dotted line after T_{swich} since the ranging sensor no longer provides any separation signal. We consider the travel time, fuel consumption and emissions for the last eight passenger vehicles after they have traveled 1.3km, and present them in Table 5-7. As we can see, though the presence of ACC_NEW significantly decreases the fuel consumption and emissions, the travel time has been

151

increased (by about 2.9 seconds). The same thing is true for heavy trucks manually driven or operated with different ACC systems.

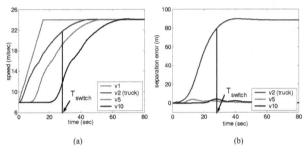

(a)   (b)

Figure 5-19: Speed limit effect: (a) speed responses and (b) separation error responses of the vehicles in string 6 (the second vehicle is an ACC truck with the new controller).

String of 10 vehicles	Manual Passenger Vehicle in the 2nd position	ACC_NEW in the 2nd position
Travel Time (sec)	68.8	71.7
Fuel (g)	774	691 (10.7%)
CO (g)	77.6	24.2 (68.8%)
HC (g)	1.84	1.06 (42.4%)
NO$_x$ (g)	3.01	1.48 (50.8%)

Table 5-7: Travel time, fuel and emission data of the 8 passenger vehicles in string 1 and 6 in the simulations with speed limit (no cut-ins).

5.5.5 Lane Change Effect

The motivation for studying lane change effects in mixed traffic with heavy trucks and passenger vehicles is based mostly on the fact that the sluggish response of the heavy truck results in a large temporary gap between the truck and the accelerating passenger vehicle in front, giving rise to multiple cut-ins from the neighboring lanes. In the simulations, it is assumed that when the spacing between the truck and its preceding

152

vehicle is large enough (larger than $s_0 + 2.8v_t$, where v_t is the truck speed), one or more passenger vehicles from the neighboring lane cut in between the two vehicles and position themselves with safe inter-vehicle spacings. The truck is considered to be equipped with the new ACC system, followed by eight passenger vehicles. We consider the situation where the passenger vehicle leading the ACC truck accelerates from 8m/sec to 20m/sec with a constant acceleration of $2m/sec^2$, and then cruises with a constant speed creating a large inter-vehicle spacing. It is assumed that three vehicles cut in at 5, 10.6 and 17.5 seconds, respectively, and with the speeds of 12.2, 18.0 and 21.0m/sec, respectively. These cut-ins are shown in Figure 5-20. We consider the travel time, fuel consumption and emissions for the last eight passenger vehicles after they have traveled 1.7km, and present them in the right column of Table 5-8. For comparison purpose, the corresponding data for the situation without cut-in vehicles are also presented in Table 5-8. From Table 5-8, we can see that the disturbance created by the cut-ins has a small negative effect on fuel consumption and emissions. However, the travel time is increased by about 4 seconds due to the cut-ins.

	Without Cut-in(s)	With Cut-in(s)
Travel Time (sec)	95.4	99.5
Fuel (g)	776	811 (-4.5%)
CO_2 (g)	2418	2531 (-4.7%)
CO (g)	24.2	24.3 (-0.4%)
HC (g)	1.08	1.11 (-2.8%)
NO_x (g)	1.36	1.40 (-2.9%)

Table 5-8: Travel time, fuel and emission data of the 8 passenger vehicles following ACC_NEW in high acceleration maneuvers, without cut-in(s) and with three cut-ins.

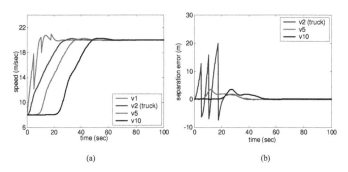

(a)   (b)

Figure 5-20: Lane change effect: (a) speed responses and (b) separation error responses of the vehicles in string 6 (the second vehicle is an ACC truck with the new controller).

From the data presented in Tables 1, 3, 5 and 7, we can see that the presence of heavy trucks in the mixed traffic can improve fuel consumption and emissions of the passenger vehicles, especially in the presence of disturbances due to high acceleration maneuvers. The heavy trucks act as filters due to their sluggish response and present to the following vehicles upstream smoother speed responses to be tracked. Furthermore, the ACC trucks lead to better fuel and emissions results than the manually driven trucks due to their smoother response. In the case of high acceleration maneuvers with oscillations, the heavy truck with the newly developed ACC controller (ACC_NEW) filters the oscillations more effectively leading to better fuel and emission results than the other ACC trucks.

In such situations another effect takes place as the sluggish response of the truck creates large inter-vehicle gaps inviting vehicles from neighboring lanes to cut in front of the truck. This effect depends on cut-in situations considered. If we consider aggressive cut-ins, for example, a passenger vehicle at a low speed cuts in just a few meters ahead of the heavy truck forcing the truck to decelerate and then speed up, the disturbance created in such scenario may have a more adverse effect on fuel economy and pollution. Furthermore cut-ins will create disturbances in the neighboring lane too with adverse effects on fuel economy and pollution for the vehicles in those lanes. Our conclusion is that while the heavy trucks with ACC or not have smooth responses and filter traffic disturbances their sluggish behavior will create large gaps inviting cut-ins which may take away any benefits their filtering response will have for the vehicles in their lane. Furthermore, the cut-ins will create additional disturbances in the neighboring lanes that will also have a negative effect on fuel economy and pollution. The travel time of the vehicle following after heavy trucks may be affected by the speed limits or cut-in vehicles.

5.6 Experimental validation

Experiments with three actual vehicles were conducted at Crows Landing test field to validate the simulation models and designed controllers. The lead vehicle was a Buick LeSabre with an automatic speed tracking system. The second vehicle was the heavy truck with different ACC systems, as shown in Figure 5-21. The tractor mass was about eight tons and the trailer mass was about seven tons. The lidar system installed on the truck was used as the ranging sensor providing separation distance and relative speed between the preceding vehicle and the truck. More detailed information about the hardware and software configurations can be found in [79]. The controllers used in the simulation section were used in the experiments. The third vehicle was a manually driven Buick LeSabre, following the truck as in normal traffic.

Figure 5-21: Photo of the experiment heavy truck at Crows Landing test filed, provided by PATH.

The experiments demonstrated that all the ACC controllers work on an actual truck in a vehicle following environment [94]. We simulate the same scenarios as in the experiments using our simulation models. In particular we consider a string of three vehicles where the first vehicle is made to generate the same speed trajectories as in the experiments, the second one is a heavy truck with one of the designed controllers, and the third one is modeled using the Pipes' model. Here we only present one set of experiment and simulation data to demonstrate the validation of the simulation results, and more data can be found in [94]. In the validation test, the first vehicle generates one speed trajectory that was used in the experiments, which is plotted with the red dotted line in Figure 5-22(a), labeled as "1st vehicle". The speed responses of the ACC truck (with the constant time headway policy) and the passenger vehicle are plotted in Figure 5-22(a) with broken lines and labeled as "2nd vehicle (ACC truck), simulation" and "3rd vehicle (Pipes' model), simulation", respectively. The corresponding experimental data are also plotted in Figure 5-22(a) with solid lines and labeled as "2nd vehicle (ACC truck), experiment" and "3rd vehicle, experiment", respectively. The engine torque responses in the simulation and experiment are plotted in Figure 5-22(b). The simulation results are very close to the experiment results. Additional validation tests supported the same conclusion: the designed vehicle following controller work on actual trucks, and the nonlinear model used for heavy trucks and the Pipes' model used to model human driver response in the longitudinal direction are valid for studying vehicle following in a mixed traffic situation. Consequently our simulation models and results can be used with confidence in studying vehicle following characteristics and effects involving more mixed traffic situation.

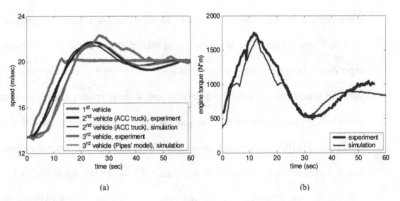

(a)   (b)

Figure 5-22: Experimental testing: (a) speed responses of the three vehicles (the second is ACC_C)
and (b) engine torque responses of ACC_C in experiment and simulation.

5.7 Conclusions

In this chapter we design, analyze and evaluate the performance of a PID type vehicle following controllers with different spacing policies. A new vehicle following controller is designed to provide better performance, which is demonstrated by simulations and emission analysis. Experiments involving actual vehicles are used to validate our simulation models at least on the microscopic level and demonstrate that the proposed vehicle following controllers work in real time and under actual driving conditions.

The sluggish dynamics of trucks whether manual or ACC due to limited acceleration and speed capabilities make their response to disturbances caused by lead passenger vehicles smooth. The vehicles following the truck are therefore presented with a smoother speed trajectory to track. This filtering effect of trucks is shown to have beneficial effects on fuel economy and pollution. The quantity of the fuel and emission benefits depends very much on the level of the disturbance and scenario of maneuvers. If the response of the truck is too sluggish relative to the speed of the lead vehicle then a large inter-vehicle spacing may be created inviting cut-ins from neighboring lanes. These cut-ins create additional disturbances with negative effects on fuel economy and pollution. Situations can be easily constructed where the benefits obtained due to the filtering effect of trucks are eliminated due to disturbances caused by possible cut-ins. Furthermore cut-ins are shown to increase travel time for the vehicles in the original traffic stream.

CHAPTER 6 : AUTOMATED CONTAINER TRANSPORT SYSTEM BETWEEN INLAND PORT AND TERMINALS

6.1 Introduction

In recent years, the global container trade has been growing at an annual rate of about 9 percent, and the corresponding U.S. rate has been around 6 percent. By 2010, it is expected that 90 percent of all liner freight will be shipped in containers. Thus every major port is expected to double or even triple its processed containers by 2020 [42]. In order to remain competitive, marine container terminals in metropolitan areas must meet the increasing demand for storage and processing capacity. Ports such as those of Los Angeles/Long Beach (LA/LB) at Southern California, which handle nearly one third of all US foreign container traffic, are under a lot of pressure to meet projected capacity demand increases in order to remain competitive and avoid traffic congestion at the terminals and contiguous areas. As potential candidates for improving the performance of container terminals and meeting the challenges of the future in marine transportation, various concepts of automated container terminals have been proposed, among which the use of automated guided vehicles (AGVs) has attracted the most studies [16, 82]. The Delta Terminal at the Port of Rotterdam has been operating AGVs for transporting containers within the terminal, while the Ports of Singapore, Thamesport, Hamburg, Kawasaki and Kaoshiung are experimenting with similar systems. Other competitive

159

concepts, including the Grid RAIL and Automated Storage/Retrieval System, are also investigated [42].

Another feasible approach to reduce the pressure of increased storage capacity demand at terminals is the use of an inland port, which acts as an intermediate storage area before the cargoes are processed for export/import. Such an inland port could be made very efficient by automating all the tasks associated with processing, scheduling, storage, and transfer of containers between the inland port and the container terminals. As an important part of such an automated system, automated trucks are employed to transport all the containers. The use of automated trucks in container transportation has a lot of benefits, such as high container throughput, 24-hour continuous operation, reduced labor cost, high reliability, high safety standards and so on. Since the automated trucks are required to transport containers between a terminal and an inland port, generally a few miles away, they will be expected to travel at much higher speeds than the AGVs operating inside container terminals. The Center of Transport Technology in the Netherlands studied a container transport system, called "Combi-Road" [68], in which each container is pulled on the semi-trailer of an unmanned vehicle, and the vehicles are electrically driven along specially designed tracks. In the US, a lot of research efforts are currently under way to study the deployment of automated commercial trucks on highways, either in platoon formations or as autonomous vehicles [13, 23, 79, 89-91, 95]. These results strongly suggest that automated commercial trucks may come into market in a near future. Despite past activities in the area of truck automation, there is currently no system that utilizes fully automated trucks at relatively high speeds, due to the human factors and liability issues.

In this chapter, we design, analyze and simulate an Automated Container Transportation system between Inland POrt and Terminals (ACTIPOT), which employs automated trucks to transfer containers between an inland port and terminals. The automated container transportation system, as shown in Figure 6-1, is composed of automated trucks, automated cranes and a supervisory controller that synchronize all the automated units inside the system. The supervisory controller contains all the information related to transportation tasks and road geometry, acquires the real time information and issues commands for all the automated units via communication devices. We design and analyze the supervisory controller using Petri Net [56], and demonstrate the overall system work in a safe and efficient manner. We briefly present the design of on-board (longitudinal and lateral) controllers for automated trucks for the completeness of the chapter. A feasible application of the ACTIPOT system is in the Long Beach area in California. In this chapter, Pier G Mega Terminal at the port of Long Beach is used as an example of the automated container terminal, and Union Pacific's ICTF (Intermodal Container Transfer Facility), located several miles north of the port of Long Beach, is used as an example of the inland port. We demonstrate that the proposed system can operate in a safe manner and achieve desired performance using microscopic simulations with the road characteristics between ICTF and Pier G. The rest of this chapter is organized as follows. In section 6.2, we introduce the basic concepts for the ACTIPOT system, and explain the design considerations and system layout. In section 6.3, we

161

briefly investigate control designs for automated trucks. The supervisory controller is designed and analyzed in section 6.4. In section 6.5, microscopic simulations are carried out to demonstrate the performance of the overall system. The conclusions are given in section 6.6.

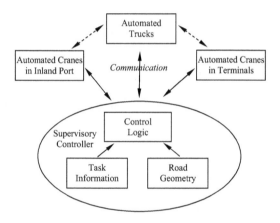

Figure 6-1: Overview of the ACTIPOT System.

6.2 ACTIPOT System

The basic components of the proposed ACTIPOT system are an inland port, terminals and an automated container transportation system. The inland port could be located a few miles away from the terminals where land of lower cost is available, and is used to temporarily store and process import/export containers. Automated trucks are used to

162

transport containers on a dedicated road inside the ACTIPOT system. The truck road may be dedicated for the automated trucks all the time or for time intervals, and the rest of the time could be used by manually driven vehicles. Inside the inland port and terminals, containers will be processed by automated cranes. As shown in Figure 6-1, the supervisory controller will synchronize the movements of all the automated units inside the transportation system via wireless communication. An automated truck employed in the ACTIPOT system will be assigned tasks such as carrying a container from the inland port, joining a platoon, speeding up to a desired speed and cruising while on the road, slowing down when entering the container terminal, positioning itself under a quay crane for unloading, then getting loaded with an imported container and driving back to the inland port, and vice versa. While being served by an automated crane, the automated truck is connected with the crane through communications, as indicated in Figure 6-1.

In this section, we present the design considerations and a feasible system layout for the ACTIPOT system, which will used in the simulations presented in sections 6.4 and 6.5. The designs of the on-board controllers for automated trucks and the supervisory controller will be presented in sections 6.3 and 6.4.

6.2.1 Design Considerations

In the design of the ACTIPOT system, we consider the following operation conditions:

1. The container terminal is able to serve ships with capacity of 8,000 Twenty-foot Equivalent Units (TEUs). It is assumed that the ships arrive every 24 hours, which requires that the service time must be strictly limited to 24 hours or less. In our design,

we further assume that the ship carries import containers up to 85% of its capacity and should be reloaded with the same number of export containers. The turnaround time for a ship with 85% load is restricted to 20 hours, so that the system is able to serve any ship within 24 hours even if the ship is fully loaded and some unexpected events take place.

2. All the import containers will be transported to the inland port before they are distributed to different destinations, and all the export cargoes will be preprocessed in the inland port before they are transferred to the container terminal. All the containers are of Forty-foot Equivalent Unit (FEU) type.

3. The maximum physical capacity of a quay crane is assumed to be 50 moves per hour in the single mode operation (i.e. either loading or unloading), and 42 moves per hour in the double mode (i.e. both loading and unloading). A variance of 15% (uniformly distributed randomness) to the maximum capacity of the quay cranes is considered, due to the uncertainties involved in the quay crane operations.

4. The maximum physical capacity of a crane in the inland port is assumed to be 60 moves per hour in the single mode. A variance of 15% is considered for this maximum capacity (uniformly distributed randomness).

5. The automated trucks are able to work 24 hours per day. No fueling or maintenance time is considered for the trucks in this study.

The above conditions are used in this chapter to demonstrate how to determine the minimum number of required quay cranes, and estimate some other specifications in the ACTIPOT system.

6.2.2 Layout of ACTIPOT System

Before we present the system layout, we need to determine how many quay cranes are required in the container terminal. A quay crane working in the dual mode transfers one export container from truck to ship and one import container from ship to truck in one moving cycle. If all the quay cranes operate at their maximum capacities, the smallest number of quay cranes required to accomplish the task proposed in section 6.2.1, N_{qc}, is given as

$$N_{qc} = \left\lceil \frac{N_{container}}{C_{qc}T_{ship}} \right\rceil \qquad (6\text{-}1)$$

where $N_{container}$ is the number of containers to be unloaded from the ship, C_{qc} is the maximum physical capacity of the quay cranes, T_{ship} is the desired ship turnaround time, and $\lceil \cdot \rceil$ is the operator that rounds up the argument to the closest integer. As explained in section 6.2.1, $N_{container}$ is also the number of containers to be loaded onto the ship. Using the specifications in section 6.2.1, we calculate that five quay cranes are required in order to load and unload a mega-ship with 3,400 containers of FEU within 20 hours.

As shown in Figure 6-2, the layout of the ACTIPOT system consists of three parts: the container terminal, the inland port and the dedicated lanes connecting the inland port with the terminal.

The container terminal, shown on the right lower box, is the place where ships are to be loaded or unloaded. When trucks are following the lane center in this area, large transient lateral errors may occur due to the large road curvatures, which could lead to a collision between two trucks traveling in opposite directions. Therefore all paths inside

165

the container terminal are designed to be uni-directional so that large transient lateral errors will not cause any problems. In this layout, five quay cranes are shown to serve the ship simultaneously, and each of the quay cranes can be accessed via the five service lanes under them. In this chapter, each truck entering the container terminal will be assigned to one quay crane with the minimum number of trucks in the service queue. At point P_1 the five service lanes merge together. To avoid collisions at this point, a time-window T_{w1} is established. When the system detects that one truck will reach P_1 at time t_1 while another truck will arrive at time t_2 and $|t_2 - t_1| < T_{w1}$, it will allow the truck closer to P_1 to pass first, while the other truck would wait until the collision possibility is eliminated. Platoon Formation (PF) is the location where automated trucks are organized to form platoons. If the PF is empty initially, the first truck that enters the PF will stay in pool 1, the second one that enters will stay in pool 2, and so on. After enough trucks have joined the platoon, the truck platoon will move towards the exit of the container terminal. In our case, we consider platoons of five trucks.

The inland port is shown in the left upper box of Figure 6-2. It has two buffers for import and export respectively and each buffer contains five cranes working in the single mode. Either of the two service lanes (labeled as 1 and 2) can access all the cranes. One platoon follows service lane 1 if available, otherwise it follows service lane 2. Similar to the container terminal, all the paths inside the inland port are uni-directional and a time-window T_{w2} is also established for the merging point P_2. There also exists a PF at the inland port to organize the trucks into platoons.

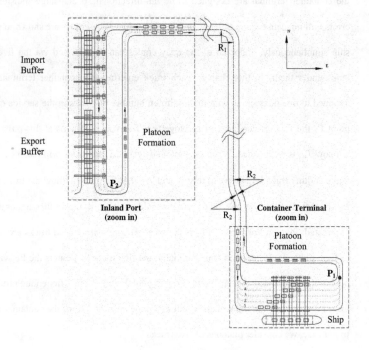

Figure 6-2: Layout for the ACTIPOT System.

The dedicated road between the inland port and the container terminal contains two uni-directional lanes of opposite directions, and automated trucks travel in platoon formations at relatively high speeds. Large road curvatures shown in Figure 6-2 are known and this knowledge is used in the design of the lateral controllers of the trucks, while the small unknown curvatures are treated as disturbances by the control system. For example, R_1=200m is known so that the curvature $1/200\text{m}^{-1}$ is taken into account by the

167

lateral controller, while the road curvature associated with R_2=1200m is treated as unknown disturbance.

6.3 Automated Trucks

In the proposed ACTIPOT system, fully automated trucks are used to transport containers on a dedicated road. These trucks are commercial heavy-duty vehicles equipped with communication devices, Differential Global Positioning System (DGPS), and on-board sensors such as Inertial Measurement Unit (IMU), radar (or lidar), wheel angle sensor et al. Through the communication devices, an automated truck receives commands from the supervisory controller to interact with the other units inside the system. The on-board sensors, together with the DGPS, provide the appropriate measurements that are used by the on-board longitudinal and lateral control systems in order to keep the truck close to the center of the lane, track desired speeds, following preceding truck and stop for loading and unloading [96].

Since vehicle control is not a contribution of this chapter, we only give a brief introduction on the truck dynamics and the design of the on-board controllers for the completeness of this work. Some similar results have been presented in various technical reports and papers [13, 23, 28, 79, 89-91]. The on-board controllers presented in this section, together with the truck dynamics, are simulated in section 6.4 to demonstrate that the use of automated trucks in the ACTIPOT system is feasible. Detailed information about the model and control parameters can be found in [96].

6.3.1 Longitudinal Control Design

Two longitudinal controllers are presented in this section, one for speed tracking and the other for vehicle following. The longitudinal truck model used for simulations is proposed by Yanakiev and Kanellakopoulos [89, 90] and has been experimentally validated [95]. It is a complicated nonlinear model characterized by a set of differential equations, algebraic relations and look-up tables. The model used for control design is [28, 89, 90].

$$\dot{v} = -a(v - v_d) + b(u - u_d) + d \tag{6-2}$$

where v is the truck speed, v_d is the desired speed, u_d is the corresponding desired fuel command, d is the modeling uncertainty, and a and b are positive constant parameters that depend on the operating point. In the vehicle following mode, the desired steady state speed is the speed of the lead vehicle v_l.

In the speed tracking mode, the longitudinal controller should regulate the vehicle speed v close to the desired speed v_d set by the supervisory controller. Here we adopt the adaptive speed tracking controller designed in Chapter 5

$$\begin{cases} u = f_u^{-1}(v_d) + k_{st,1}e_v + k_{st,2} + k_{st,3}\dot{v}_d \\ \dot{k}_{st,1} = \text{Proj}\{\gamma_{st,1}e_v^2\}, \dot{k}_{st,2} = \text{Proj}\{\gamma_{st,2}e_v\}, \dot{k}_{st,3} = \text{Proj}\{\gamma_{st,3}e_v\dot{v}_d\} \end{cases} \tag{6-3}$$

where $e_v = v_d - v$ is the speed error, $k_{st,i}$ $(i=1,2,3)$ are the control gains, $\gamma_{st,i}$ $(i=1,2,3)$ are positive design parameters, and $\text{Proj}\{\cdot\}$ is the projection function, can stabilize the closed-loop system. If v_d and d are constants, then e_v converges to zero as time goes to infinite. Using (6-3) the fuel command is issued when u is positive, while the brake is

169

activated when $u < -u_0$ (u_0 is a positive constant). Otherwise, the brake system is inactive and the fuel system is operating as in idle speed.

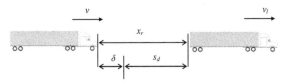

Figure 6-3: Trucks in the vehicle following mode.

As shown in Figure 6-3, in the vehicle following control mode, the longitudinal controller should regulate the truck speed v towards the speed of the lead truck v_l and keep the intervehicle spacing x_r close to the desired spacing s_d. Let us define the relative speed as $v_r = v_l - v$ and the separation error as $\delta = x_r - s_d$. With the time headway policy, s_d is given by $s_d = s_0 + hv$, where s_0 is a fixed safety spacing and h is the time headway. In [91], a variable time headway is proposed as

$$h = \text{sat}(h_0 - c_h v_r) \qquad\qquad (6\text{-}4)$$

where h_0 and c_h are positive design parameters, and sat($\cdot$) is the saturation function with an upper bound 1 and lower bound 0. This variable time headway is adopted in our study since it has been shown to be able to provide safe vehicle following with tight intervehicle spacing. Here we adopt the vehicle following controller from Chapter 4

$$\begin{cases} u = k_{vf,1}v_r + k_{vf,2}\delta + k_{vf,3} \\ \dot{k}_{vf,1} = \text{Proj}\left\{\gamma_{vf,1}v_r\left[(p_1v_r+\delta)+(v_r+a_mkp_1\delta+a_m\delta)H\right]\right\} \\ \dot{k}_{vf,2} = \text{Proj}\left\{\gamma_{vf,2}\delta\left[(p_1v_r+\delta)+(v_r+a_mkp_1\delta+a_m\delta)H\right]\right\} \\ \dot{k}_{vf,3} = \text{Proj}\left\{\gamma_{vf,3}\left[(p_1v_r+\delta)+(v_r+a_mkp_1\delta+a_m\delta)H\right]\right\} \end{cases} \qquad (6\text{-}5)$$

where $k_{vf,1}$, $k_{vf,2}$ and $k_{vf,3}$ are variable control gains, a_m, k and $\gamma_{vf,i}$ (i=1,2,3) are positive design parameters, and $H = \partial s_d / \partial v$. The adaptive controller (6-5) can stabilize the closed-loop system if the design parameters are chosen such that

$$\begin{cases} a_mp_1 > 1 \\ \dfrac{4p_1k}{a_m+a_mkp_1-k} > \sup H \end{cases} \qquad (6\text{-}6)$$

where p_1 is a constant. Furthermore, if v_l and d are constants, then $v_r,\delta \to 0$ as $t \to \infty$.

In the simulations in section 6.4.3, the parameters used in the variable time headway are chosen as s_0=3m, h_0=0.1, and c_h=0.2 [91]. We choose $a_m = 0.5$, $p_1 = 10$ and k=1.0 so that (6-6) holds when the lead truck operates under the maximum speed of 30m/s. The other control parameters are chosen based on simulations to achieve good vehicle following performance and are not presented here. According to (6-5), fuel and brake commands are issued in the same way as in the speed tracking control.

6.3.2 Lateral Control Design

We design the lateral controller to generate steering commands by using the McFarlane and Glover loop-shaping method [46], which has been implemented for lateral control of heavy-duty vehicles [23]. We use the same technique with different nominal values for trailer mass, longitudinal speed and road adhesion coefficient, which are appropriate for our specific applications.

The lateral model with respect to the road reference frame has the form [13]

$$M\ddot{q}_r + D\dot{q}_r + Kq_r = F\theta_w + E_1\dot{\varepsilon}_d + E_2\ddot{\varepsilon}_d \tag{6-7}$$

where $q_r = [y_r \quad \varepsilon_r \quad \varepsilon_f]^T$, y_r is the lateral displacement of tractor's center of gravity (CG) with respect to the road center line, ε_r is the yaw angle of the tractor relative to the road center line, ε_f is the relative yaw angle between the tractor and the semi-trailer (see Figure 6-4), θ_w is the front wheel angle and $\dot{\varepsilon}_d$ and $\ddot{\varepsilon}_d$ are road curvature characteristics. The matrices M, D, K, F, E_1 and E_2 are related to the truck characteristics [96]. The steering system is approximated as a first-order system

$$\frac{\theta_w(s)}{\theta_s(s)} = \frac{1}{0.08s + 1} \tag{6-8}$$

where θ_s is the steering angle. The transfer function from θ_s to y_r obtained from (6-7) and (6-8) contains a pair of poorly-damped zeros when the longitudinal velocity is high, which makes the lateral control difficult if the lateral error at the tractor's CG is the only signal used for feedback. This problem is solved by introducing a *look-ahead distance d_s* [23]. As shown in Figure 6-4, it is assumed that a virtual sensor is placed at distance d_s ahead of the tractor's CG, and its measurement y_s is used for feedback. With

$$y_s = y_r + d_s\varepsilon_r \tag{6-9}$$

we can get $G_o(s)$, the transfer function from θ_s to y_s. Model uncertainties in $G_o(s)$ may be due to variations in trailer mass m_2, longitudinal speed v and road adhesion coefficient μ.

172

The McFarlane and Glover loop-shaping method is applied here to handle the uncertainties. In our control design, d_s is selected to be 5m, and the nominal values for the variable parameters are m_2=15,000kg, v=20.1m/s and μ=0.8.

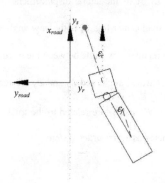

Figure 6-4: A truck in the road reference frame.

6.4 Supervisory Controller

In this section, we design and analyze the supervisory controller that dictates and synchronizes the movements of the trucks and cranes. The supervisory controller assigns new tasks to cranes, checks truck positions, generates proper speed trajectories and selects appropriate longitudinal/lateral actions for the trucks under different situations. As shown in Figure 6-5, the supervisory controller is composed of two units: the Information Center and the Control Logic. All necessary information, such as path information, ship arrival and departure times, tasks to be performed and so on, are stored in advance in the

Information Center. Every unit in the ACTIPOT system provides its updated status to the Information Center by direct communication. The Control Logic requires information from the Information Center and instructs all the units in the ACTIPOT system. We also investigate the performance of a platoon of trucks under the guidance of the supervisory controller through simulations.

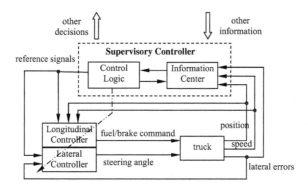

Figure 6-5: Interaction between the supervisory controller and a truck.

6.4.1 Control Logic

In the ACTIPOT system, a platoon with import containers slows down as it enters the inland port, checks which service lane is available, then splits so that the trucks operate in the individual mode. Individual trucks position themselves under the assigned cranes in the Import/Export Buffer to get unloaded and loaded, move toward PF and wait there until a platoon is formed. The newly formed platoon moves towards the exit of the inland port, speeds up to a desired speed while cruising on the dedicated lanes, slows down to enter the container terminal, allows splitting of the trucks in such a way that each truck will be assigned to one quay crane with the minimum number of trucks in the service queue. After being unloaded, a truck picks up another container and moves towards the PF to form the next platoon. The new platoon moves back to the inland port and the same process is repeated for all platoons.

In the ACTIPOT system, the basic requirement for the supervisory controller is to guarantee operations with no collision, no congestion and good performance. There are three possibilities for two trucks to collide in the ACTIPOT system: (i) when two trucks are merging simultaneously at the point $\mathbf{P_1}$ or $\mathbf{P_2}$ in the layout shown in Figure 6-2, (ii) when the trucks are moving on different lanes, but they may interfere with each other due to large lateral errors and (iii) when two trucks are traveling on the same lane and the intervehicle spacing between them is unsafe. The first possibility has been automatically excluded by establishing the two time windows T_{w1} and T_{w2} so that two trucks will never merge at the same time. The second possibility is eliminated too by designing robust lateral controllers that always keep the trucks very close to the center of the lane. The collision between two trucks traveling on the same lane is avoided by using intervehicle spacing that is safe under the worst stopping and accelerating conditions. This consideration leads to the following intervehicle spacing

$$S_{safe} = \begin{cases} S_{\min}, & v \le 3.6\text{m/s} \\ \eta S_{stop}, & \text{otherwise} \end{cases} \tag{6-10}$$

175

where S_{safe} is the safety spacing, S_{stop} is the stopping distance obtained based on the simulated characteristics of the truck, the constant η is set as 1.2 in our simulations, and S_{min} is set equal to η times the stopping distance for a truck traveling at 3.6m/s (about 8 miles/hour, which is the nominal speed for trucks in the inland port and terminals). Once a truck detects that there is another truck ahead in the same lane and the distance between them is less than S_{safe}, it will decelerate until a safe intervehicle spacing is reached. Once the supervisory controller properly selects the longitudinal controllers and speed trajectories, the system safety can always be guaranteed.

The layout shown in Figure 6-2 and operations are designed for congestion-free environment. There will be no deadlock in the system if the supervisory controller has no deadlock. It is clear that the longitudinal behavior of the automated trucks is an important issue in the ACTIPOT system, and that the longitudinal control logic is the most critical component of the overall control logic. According to the longitudinal speed command, a truck in the individual mode is considered to have four states: acceleration, deceleration, cruise or stop. In the acceleration state, the supervisory controller generates a smooth increasing speed trajectory to be followed. The desired acceleration is varies from 0.5m/s^2 to 0.2m/s^2 depending on the truck load and speed. In the deceleration state, the supervisory controller generates a decreasing speed trajectory to be followed, which has an acceleration of -2m/s^2. In the cruising case, the truck follows a constant speed. In the stop state, the brake is always on so that the truck keeps still all the time. For a platoon, the leading truck is considered as operating in the individual mode by tracking assigned

176

speed trajectories, but the following trucks are considered as operating in the vehicle following mode.

6.4.2 Petri Net Modeling and Analysis

Petri Net is a graphic and mathematical modeling tool applicable to many systems. In this chapter we use it to model the supervisory controller since the graphic presentation makes the controller easy to understand. The Petri Net model of the supervisory controller consists of two sub-modules, one for trucks and one for cranes. We investigate the liveness and safeness properties of the modules individually and for the overall system. The system is dead-lock free and will not lead to conflicted decisions if it is live and safe. The basic definitions and theory of Petri Nets can be found in references such as [56].

6.4.2.1 Crane Module

As shown in Figure 6-6(a), the control logic for a single-mode crane in the Import Buffer has four places:

    1. *crane_idling*: no job is assigned to the crane, or the crane is ready to serve a truck but the truck has not arrived. A crane in this mode will keep idling until a job becomes available.

    2. *unload_truck*: the crane unloads a container from a truck.

    3. *move_container*: the crane moves the container towards the Import Buffer and stacks it.

    4. *move_back*: the crane moves back for the next available job .

The transitions in this model are obvious, and the associated description is omitted. Similarly, the control logic for a single-mode crane in the Export Buffer is modeled in Figure 6-6(b). It also has four places:

1. *crane_idling*: this is the same as above.

2. *load_truck*: the crane loads a container onto a truck.

3. *move_back*: the crane moves towards the Export Buffer for the next container if available.

4: *move_container*: the crane moves a container from the Export Buffer to serve the arriving truck.

The control logic for a dual-mode quay crane is modeled in Figure 6-6(c), which has two places:

1. *crane_idling*: no truck is assigned or the crane is waiting for the next coming truck.

2. *serve_truck*: the crane unloads an export container from the truck, moves it to the ship, stacks it, moves back with an import container and loads it onto the truck.

It is easy to see that each of the three crane modules is a strongly connected State Machine (SM) with only one token. Therefore, they are live, safe [56]. It can also be seen that a live and safe SM (N, M_0) is reversible, since a token that leaves a place can always go back to the same place.

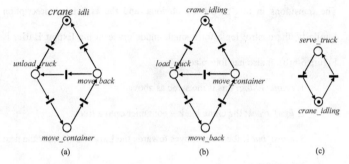

Figure 6-6: Petri Net modules for (a) a single-mode crane in the Import Buffer, (b) a single-mode crane in the Export Buffer and (c) a dual-mode quay crane in the container terminal.

6.4.2.2 Truck Module

The control logic for an automated truck can be divided into three parts. As shown in Figure 6-7, they are "system check", "safety check" and "decision control". In fact, the control decisions, such as which on-board controllers to be used and what kind of speed commands to be generated, are all decided by the key part "decision control". The first two parts are employed to assist "decision control", and will not generate any control decision directly. In other words, they can be implicitly included into the transitions of "decision control" as we will discuss later. The first part "system check" is used to check the functional status of the on-board systems and it has three places:

    1. *system_OK*: A token is put in this place when all the on-board systems operate properly.

179

2. *system_check*: It is checking the on-board systems continuously, until all the containers have been transported. It is assumed that this check is almost instantaneous and does not introduce significant time delays.

3. *system_failure*: A token is put in this place once a system failure is detected.

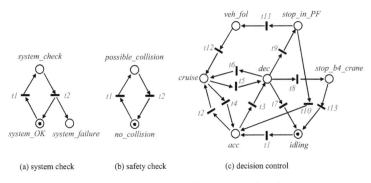

(a) system check   (b) safety check   (c) decision control

Figure 6-7: Petri Net module for an automated truck.

The second part "safety check" incorporates the safety policy in (6-10) into the supervisory controller, and it has two places:

1. *no_collision*: A token in this place indicates the safety policy is not violated.

2. *possible_collision*: A token in this place indicates that the truck gets too close to another truck ahead and collision possibility exists if the current speed is maintained.

180

The third part, which is also the key part of the supervisory controller, is used to select appropriate controllers for trucks and provide reference signals if necessary. As shown in Figure 6-7(c), it has seven places and thirteen transitions. The seven places in fact represent seven truck working states:

1. *idling*: The truck stays still in this state. This happens when possible collision exists ahead or no job is assigned. The brake is always on during this state.

2. *acc*: The truck tracks a desired increasing speed trajectory and the longitudinal controller in (6-3) is engaged.

3. *cruise*: The truck tracks a constant speed and the controller in (6-3) is engaged.

4. *dec*: The truck tracks a decreasing speed trajectory and the controller in (6-3) is engaged.

5. *stop_b4_crane*: The truck stops before the assigned crane and waits until the service is complete. The brake is always on during this state.

6. *stop_in_PF*: The truck stops in PF and waits until the platoon is formed. The brake is always on during this state.

7. *veh_fol*: The truck is part of a platoon, but not a leader. In this case, it follows the preceding truck. The longitudinal controller in (6-5) is engaged.

The thirteen transitions that represent different logic operations are:

1. *t1*: If the truck stops to avoid a possible collision, and the possibility of this collision has been eliminated, then the truck will begin to accelerate.

2. *t2*: If the truck has reached the speed limit, then it will track this limit speed.

3. *t3*: If the truck detects a collision possibility during acceleration, it will slow down.

4. *t4*: If the truck has been cruising at a speed below the speed limit for some reason (we will come to this point at *t6*), and there is no collision possibility, then it will speed up.

5. *t5*: If the truck has detected a possible collision ahead or it needs to slow down to enter the service destination, it will decelerate until the collision possibility vanishes or it reaches the destination.

6. *t6*: If the truck decelerates because of possible collision and this collision possibility has vanished, then it will cruise at the current speed for a few seconds. During this cruising period, if the collision possibility reappears, then *t5* will take the truck back to *dec*. Otherwise, *t4* will transition the truck to the *acc*. There is no direct transition from *dec* to *acc*, because it may cause chattering.

7. *t7*: If a collision possibility exists during *dec*, then the truck will come to a complete stop.

8. *t8*: When the truck arrives at the service destination point, it will stop and wait there until it is served.

9. *t9*: When the truck enters the PF point, it will stop and wait until the platoon is formed.

10. *t10*: If the truck is the leading truck in a formed platoon, then it can transition to *acc* in a similar fashion as a truck in the autonomous mode.

11. *t11*: When the truck is within a formed platoon, then it will enter *veh_fol*.

12. *t12*: When the truck separates from a platoon, it will enter *cruise*.

13. *t13*: After the truck is served by a crane, it will enter *idling*.

As mentioned before, the sub-module "system check" has been incorporated into the transitions in "decision control". A token inside *system_OK* means any transition in "decision control" is executable, while a token inside *system_failure* forbids all transitions except *t3*, *t5* and *t7*. Thus the truck must come to a complete stop and wait to be towed away. Since the truck with failure will be removed from the system eventually, it can only affect system performance but not system liveness. Similarly, in "safety check", a token inside *possible_collision* will disable all the transitions except *t3*, *t5* and *t7*. But when the collision possibility disappears, the token will move back to *no_collision*, which will make all transitions valid again. In our analysis, we refer to the part "decision control" as the main control logic for trucks.

It is easy to see that the truck module is a strongly connected state machine with only one token. Hence it is live and safe.

6.4.2.3 Supervisory Controller

Although the truck module and the crane modules can be modeled and analyzed independently, they are not completely isolated. There is a "dynamic transition" between a truck and a crane, which dynamically links trucks and cranes together. This link exists only when one truck is under the service of the assigned crane, as shown in Figure 6-8. After the service is completed, this connection will automatically disappear. Such a "dynamic transition" will not appear until the truck is assigned to another crane. The following lemma tells that the supervisory controller is live and safe.

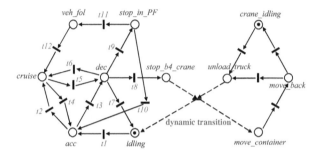

Figure 6-8: Dynamic transition between an automated truck and a single-mode crane.

Lemma 6-1: With the dynamic transition shown in Figure 6-8, the supervisory controller modeled with Petri Nets involving automated trucks and cranes is live and safe.

Proof: Suppose at some time point, the overall system is not live, i.e. there exists at least one deadlock in the overall system. This deadlock would correspond to one of the following three cases:

1. A token inside *system_failure* causes a deadlock for the truck. In this case, as soon as the system failure is detected, the truck comes to a stop and keeps idling. It may block the road in the ACTIPOT system and cause deadlock. However, the system liveness will be recovered once the failed truck is removed from the system.

2. A deadlock exists inside the "decision control" module for an active truck. From the previous analysis, we know that this could only happen when no crane is assigned to serve that truck. However, in our design, this is not possible as each truck entering the container terminal or inland port will be assigned to one and only one crane, and each truck can follow one and only one assigned lane.

3. There is a deadlock in one crane module. From Figure 6-6, we can see that this deadlock may happen when there is no truck assigned to that crane. Since we have designed the service lanes so that any crane is accessible to any truck in the system, no truck assigned to the crane means the crane has broken down or the path to it is blocked for emergency. However, it will not affect the liveness of the whole system since the other cranes still work. Once we properly revise the dispatching rule so that all the trucks are assigned only to the active cranes, the system is still live. This point is verified using microscopic simulations, where a crane that cannot work properly is simulated.

Given the above discussions we can conclude that the overall system is live. The control logic in the supervisory controller can be viewed as a collection of relatively independent sub-modules connected by the dynamical transitions. Furthermore, since a token in the overall system is always inside one sub-module, the Petri Net model for the overall system is safe.

Remark 6-1: In this work, we consider all the containers are temporally storied and processed in the inland port. Hence optimization of the traffic is not an issue for the considered system. In the case that some containers are processed and stored in the terminals, the results presented in [42] and [37] can be applied.

6.4.3 Illustrative Simulations

Using the developed supervisory, longitudinal and lateral controllers, we study the behavior of one platoon in the ACTIPOT system, which has the layout in Figure 6-2. In the simulations, the length of dedicated road is about 4.3 miles, which is the distance between ICTF and Pier G. The platoon, composed of 5 trucks (simulated with the nonlinear models) loaded with FEU containers of 15 tons, speeds up to 20.1m/s (45 miles per hour), cruises with this speed towards the container terminal, and slows down to 3.6 m/s (8 miles per hour) to enter the container terminal. The speed, relative speed, separation distance and separation error responses are shown in Figure 6-9. The speed, relative speed and separation distance vehicle responses indicate that the platoon can be viewed as a single moving unit. As the lead truck cruises at a constant speed, the relative speeds and separation errors within the platoon go to zero. Figure 6-9(c) shows that the separation distance between any two adjacent trucks is larger than 2.8m even in the worst case, indicating a collision-free operation. Figure 6-10(a) shows the steering angle responses of the five trucks, and Figures 6-12(b) to 6-12(d) show lateral displacements from the lane center at the tractor front axis, rear axis and trailer rear axis. The lateral displacements are always small, which means that the trucks are kept within the

186

dedicated lanes so that no collision could happen between two units operating on different lanes.

The simulation results demonstrate that there is no collision within the platoon under all possible maneuvers, and all trucks exhibit satisfactory speed responses. At the same time, all the trucks are kept close enough to the center of the dedicated lanes. After a truck is released from a platoon, it follows the assigned service lane, cruises at a speed of 3.6 m/s towards the assigned crane, stops under the crane and waits until it is served. It then moves to the PF point and waits for a platoon formation. The simulation results are not presented here, and interested readers are referred to [96].

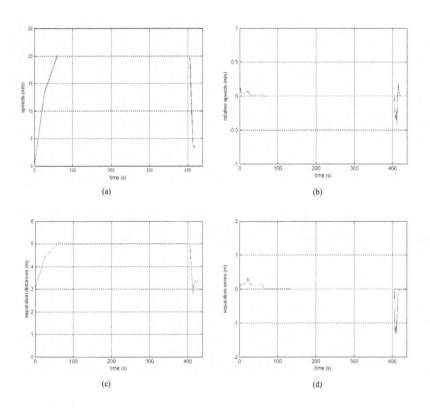

Figure 6-9: (a) Speed, (b) relative speed, (c) separation distance and (d) separation error responses within the platoon.

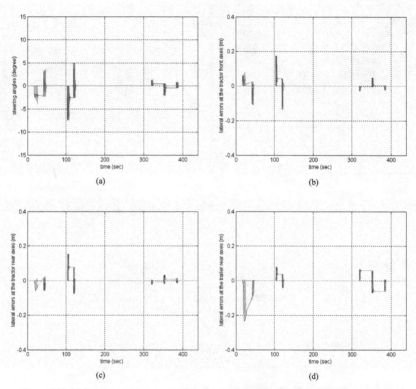

Figure 6-10: (a) Steering angle responses of the five trucks, and lateral errors at (b) the tractor front axis, (c) rear axis and (d) trailer rear axis of the five trucks.

6.5 Evaluation of ACTIPOT System

We have demonstrated in section 6.4.3 that the trucks can always be kept within the dedicated lanes without colliding or getting off the assigned lane. In this section, we investigate the performance of the proposed ACTIPOT system in different situations with

189

a large number of trucks. In this case, the dynamics of the automated trucks have to be highly simplified to reduce the simulation complexity so that the simulation is computational feasible on a personal computer. Since the lateral behavior of a truck will not lead to collisions and has little effect on the truck traveling time, the lateral truck dynamics are neglected. Another simplification is to approximate the longitudinal behavior of an automated truck in the individual mode with a simple time delay

$$v(t) = v_d(t - \tau) \tag{6-11}$$

where $v(t)$ is the longitudinal speed of the truck, $v_d(t)$ is the desired speed and τ is a small time delay. An automated truck in the vehicle following mode is assumed to keep the intervehicle spacing close to the desired spacing. We use Matlab/Stateflow for the following microscopic simulations. The simplified truck model in (6-11) and the supervisory control logic presented in section 6.4 are used.

Before we present the simulation results, some key definitions are introduced. Let us define the truck turnaround time, T_{truck}, as the average time for one truck to transport one container from the inland port to the container terminal and transport another container back to the inland port when there is no traffic congestion. The efficiency of the ith quay crane can be evaluated by its busy rate $BR_{qc}(i)$, which is defined as

$$BR_{qc}(i) = \frac{Busy\ Period\,(i)}{Busy\ Period\,(i) + Idle\ Period\,(i)} \tag{6-12}$$

where *Busy Period*(i) represents the total time that the ith quay crane is engaged for loading/unloading trucks, and *Idle Period*(i) is the total time that the ith quay crane is not engaged in any job. The average busy rate of the quay cranes, BR_{qc}, is defined as

190

$$BR_{qc} = \frac{1}{N_{qc}} \sum_{i=1}^{N_{qc}} BR_{qc}(i) \tag{6-13}$$

The average busy rate of the cranes in the inland port, BR_{pc}, is defined in the same manner. The efficiency of the trucks can be similarly evaluated by their average busy rate, BR_{truck}, defined as

$$BR_{truck} = \frac{N_{container} T_{truck}}{N_{truck} T_{system}} \tag{6-14}$$

where T_{system} is the total time for the ACTIPOT system to accomplish the assigned transportation task. If the ACTIPOT system is designed properly, the busy rates of the automated cranes and trucks should be kept close to 1 and the ship turnaround time should be within the desired time window.

One question arises: how many automated trucks are required to keep the efficiency of the ACTIPOT system high? Given the desired ship turnaround time T_{ship}, the minimum number of trucks required to transport all containers within the time window is obtained as

$$\underline{N}_{truck} = \left\lceil \frac{N_{container} T_{truck}}{T_{ship}} \right\rceil \tag{6-15}$$

However, if too many trucks are employed in the system, then terrible traffic congestion will show up and bring down the system efficiency. Hence the maximum number of trucks that should be employed is

$$\overline{N}_{truck} = \left\lceil N_{qc} C_{qc} T_{truck} \right\rceil \tag{6-16}$$

which is just enough to keep the efficiency of the quay cranes close to 1. Since the number of the quay cranes, N_{qc}, is determined by (6-1), it is easy to see that $\overline{N}_{truck} \geq \underline{N}_{truck}$.

6.5.1 Case 1

In this case, the layout of the ACTIPOT system shown in Figure 6-2 is simulated with the system description given in section 6.2.1. The layout in Figure 6-2 corresponds to the area between the ICTF and Pier G in the Long Beach area. With the illustrative simulation results shown in section 6.4.3, we know the truck turnaround time T_{truck} is about 1464 seconds [96]. Hence we can predict that the system efficiency should be high when the number of trucks stays between $\underline{N}_{truck} = 70$ $\overline{N}_{truck} = 86$. The simulation results are presented in Figure 6-11. When 80 trucks are employed in the ACTIPOT system, the ship turnaround time is close to the minimum value and the quay crane busy rate is kept close to its maximum value. At the same time, the traffic congestion is not serious since the truck busy rate is close to 1. By increasing the number of trucks beyond 80, the ship turnaround time does not decrease significantly, but the traffic congestion becomes more and more serious, which means the system efficiency is decreasing. When 110 trucks are used, the truck busy rate gets down to 0.7, which means there are too many unnecessary stop-and-go motions. When decreasing the number of trucks below 80, the truck busy rate goes up, but the ship turnaround time increases, too. When 70 trucks are used, the ship turnaround time is about 20 hours, which merely meets the desired time. Any malfunctions in the ACTIPOT system may shift the turnaround time outside the acceptable region. It should be noticed that BR_{qc} cannot get too close to 1 when the quay cranes operate in the dual mode, because it takes some time for a truck to position itself under the assigned quay crane, and the crane in the dual mode has to idle during that period.

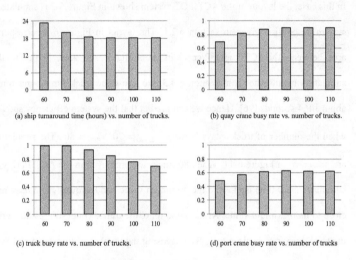

(a) ship turnaround time (hours) vs. number of trucks.

(b) quay crane busy rate vs. number of trucks.

(c) truck busy rate vs. number of trucks.

(d) port crane busy rate vs. number of trucks

Figure 6-11: Simulation results for case 1.

With the truck number fixed at 80, simulations are performed for ships with different load ratios (load ratio: the ratio of shipload to ship capacity) and the results are given in Figure 6-12. It is clear that there is a linear relation between ship turnaround time and the load ratio, which indicates that the system performance is insensitive to crane uncertainties. Furthermore, the ship turnaround time is much less than 24 hours for a fully loaded ship.

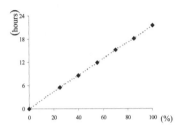

Figure 6-12: Ship turnaround time vs. shipload ratio.

6.5.2 Case 2

In the second case, the simulation layout is the same as that in Figure 6-2 except the inland port. The layout for the inland port is modified to be the same as that for the container terminal shown in Figure 6-2. In such a case, the ACTIPOT system is almost symmetric. With the truck turnaround time 1590 seconds [96], we can predict that the system efficiency should be high when the number of trucks stays between $\underline{N}_{truck} = 76$ and $\overline{N}_{truck} = 93$. The simulation results in Figure 6-13 show that 80 or more trucks are needed to accomplish the transportation task within 20 hours, and the optimum number of trucks does lie between $\underline{N}_{truck}$ and $\overline{N}_{truck}$.

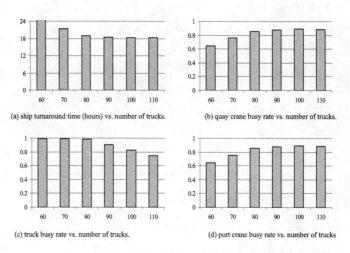

(a) ship turnaround time (hours) vs. number of trucks.

(b) quay crane busy rate vs. number of trucks.

(c) truck busy rate vs. number of trucks.

(d) port crane busy rate vs. number of trucks

Figure 6-13: Simulation results for case 2.

In all the simulations, no collision situation has ever been encountered. The simulation results demonstrate that the ACTIPOT system operates as designed during normal operations when the truck number is properly chosen.

6.6 Conclusions

In this study we propose the new concept referred to as the ACTIPOT system, in which fully automated trucks are utilized to transport containers between inland port and terminals. We design, analyze and simulate the ACTIPOT system with emphasis on the supervisory controller that synchronizes all the movements inside the system. It is demonstrated that each subsystem in the ACTIPOT system operates in a satisfactory manner and the overall performance is what is expected.

Our preliminary study indicates that the ACTIPOT system is feasible and could operate in an efficient manner. The issues that require further investigation are cost analysis and effectiveness as well as acceptance by terminals and other stakeholders. Furthermore technical issues such as particular choices of sensors, actuators, equipment based on cost, reliability and performance considerations need to be addressed by performing actual experiments and additional studies. Another important issue is the location of the ACTIPOT system and the availability of land for an inland port and of dedicated lanes to connect the inland port with the terminals. Since our approach focuses on the deployment of automated trucks in a controlled environment where humans are not present, the use of automated trucks is free of the human factors and liability issues. Hence truck automation is strongly feasible and it will be acceptable in the ACTIPOT environment provided its benefits can be established.

CHAPTER 7 : SUMMARY

In the past fifty years, adaptive control theory has reached a relatively high level of maturity. Various identifier or non-identifier based adaptive schemes have been developed with well established stability results, and many successful applications have been reported. However, applications of adaptive control in safety sensitive systems are still very limited. Its complexity in stability analysis and lack of guaranteed performance make practitioners reluctant to implement an adaptive controller in safety sensitive systems. On the other hand, the popularly accepted control laws, such as robust controllers, do not account for unpredictable changes in the plant and may deteriorate performance when system failure occurs or the parameters change drastically.

The purpose of this dissertation to investigate via theoretical analysis and practical applications the design of adaptive control schemes which are safe to implement in environments where safety is a constraint and performance guarantees are required. In this dissertation, we first investigate a non-identifier based adaptive scheme, which can be viewed as a supervisory control problem. It is shown that the supervisor can be designed so that the closed-loop system is guaranteed to be stable as long as the candidate controller set carries certain properties and there is one finite-gain stabilizing controller available. In the special case where all the sub-systems are linear time invariant, this adaptive scheme guarantees system stability without imposing any assumptions, such as stable or minimum-phase, on either the candidate controllers or the plant. This supervisory control scheme is modified to develop a safe adaptive controller in the model

197

reference control case. The safe adaptive controller is composed of one supervisory controller and two candidate model reference controllers: one is non-adaptive stabilizing while the other is adaptive. The supervisory controller is designed to guide the switching process among the two candidate controllers in a safe manner, and it switches on the adaptive controller only when it can improve system performance. It has been shown that the proposed safe adaptive controller guarantees system stability even if the adaptive controller is destabilizing. The simulation results demonstrate that it can achieve superior performance than either of the two candidate controllers. Extension of the safe adaptive controller in the case of pole placement control design is still under investigation.

In this dissertation, we also investigate how conventional robust adaptive controllers can be implemented in intelligent transportation systems to achieve desired performance. In particular, we consider how to implement adaptive control in the vehicle following control problem. We have designed an adaptive vehicle following control system that can be implemented with a general nonlinear time headway. It is shown that the designed control system guarantees system stability and it regulates the speed and separation errors towards zero when the lead vehicle is at a constant speed. The performance of the designed adaptive vehicle following controller is demonstrated by the simulations using a validated nonlinear passenger vehicle model. With the considerations of traffic flow and fuel economy, another vehicle following controller has also been developed in this dissertation, which is formed by a nonlinear reference speed generator and a robust adaptive speed tracking controller. It has been demonstrated that this special adaptive vehicle following controller has better properties than existing ones with respect to performance and impact on fuel economy and pollution during traffic disturbances. Finally, the developed vehicle following controllers are adopted to design fully automated trucks in the ACTIPOT system. As demonstrated by simulation results, the fully automated trucks can operate in a safe and efficient manner in the ACTIPOT system and achieve desired performance. All the practical applications investigated in this dissertation have indicated that the conventional robust adaptive controllers can be implemented safely and achieve desired performance with proper designs.

BIBLIOGRAPHY

[1]   T. Agnoloni and E. Mosca, "Controller falsification based on multiple models," *International Journal of Adaptive Control and Signal Processing*, vol. 17, pp. 163-177, 2003.

[2]   B. D. O. Anderson, T. Brinsmead, D. Liberzon, and A. S. Morse, "Multiple model adaptive control with safe switching," *International Journal of Adaptive Control and Signal Processing*, vol. 15, pp. 445-470, 2001.

[3]   N. Andreiev, "A process controller that adapts to signal and process conditions," *Control Engineering*, vol. 24, pp. 38-40, 1977.

[4]   J. A. Aseltine, A. R. Mancini, and C. W. Sartune, "A survey of adaptive control systems," *IRE Transactions on Automatic Control*, vol. 3, pp. 102-108, 1958.

[5]   M. Athans, S. Fekri, and A. Pascoal, "Issues on robust adaptive feedback control," *Proceedings of the16th IFAC World Congress*, 2005.

[6]   H. S. Bae and J. C. Gerdes, "Command modification using input shaping for automated highway systems with heavy trucks," *Proceedings of the 2003 American Control Conference*, vol. 1, pp. 54-59 vol.1, 2003.

[7]   M. Bando, K. Hasebe, A. Nakayama, A. Shibata, and Y. Sugiyama, "Dynamical model of traffic congestion and numerical-simulation," *Physical Review E*, vol. 51, pp. 1035-1042, 1995.

[8]   M. J. Barth, *User's guide: Comprehensive Modal Emissions Model, version 2.0*: University of California, Riverside, 2000.

[9]   J. Bishop and W. Stevens, "Results of precursor systems analyses of automated highway systems," *Proceedings of First World Congress on Applications of Transport Telematics and Intelligent Vehicle*, 1997.

[10]   A. Bose and P. A. Ioannou, "Analysis of traffic flow with mixed manual and intelligent cruise control vehicles: Theory and experiments," California PATH Research Report UCB-ITS-PRR-2001-13, 2001.

[11]   A. Bose and P. A. Ioannou, "Analysis of traffic flow with mixed manual and semiautomated vehicles," *IEEE Transactions on Intelligent Transportation Systems*, vol. 4, pp. 173-188, 2003.

[12] F. Broqua, G. Lerner, V. Mauro, and E. Morello, "Cooperative driving: basic concepts and a first assessment of intelligent cruise control strategies," *Proceedings of the DRIVE Conference on Advanced Telematics in Road Guidance*, pp. 908-929, 1991.

[13] C. Chen and M. Tomizuka, "Dynamic modeling of articulated vehicles for automated highway systems," *Proceedings of the 1995 American Control Conference*, vol. 1, pp. 653-657 vol.1, 1995.

[14] N. E. Cotter, "The Stone-Weierstrass theorem and its application to neural networks," *IEEE Transactions on Neural Networks*, vol. 1, pp. 290-295, 1990.

[15] B. Egardt, "Stability of Adaptive Controllers," in *Lecture Notes in Control and Information Sciences*, vol. 20: Springer-Verlag, Berlin, 1979.

[16] J. J. M. Evers and S. A. J. Koppers, "Automated guided vehicle traffic control at a container terminal," *Transportation Research Part a-Policy and Practice*, vol. 30, pp. 21-34, 1996.

[17] B. Fidan, E. B. Kosmatopoulos, and P. A. Ioannou, "A switching controller for multivariable LTI systems with known and unknown parameters," *Proceedings of the 41st IEEE Conference on Decision and Control*, vol. 4, pp. 4688-4693, 2002.

[18] M. Fu and B. Barmish, "Adaptive stabilization of linear systems via switching control," *IEEE Transactions on Automatic Control*, vol. 31, pp. 1097-1103, 1986.

[19] S. Fujii, J. P. Hespanha, and A. S. Morse, "Supervisory control of families of noise suppressing controllers," *Proceedings of the 37th IEEE Conference on Decision and Control*, vol. 2, pp. 1641-1646, 1998.

[20] G. C. Goodwin, P. J. Ramadge, and P. E. Caines, "Discrete-time multivariable adaptive-control," *IEEE Transactions on Automatic Control*, vol. 25, pp. 449-456, 1980.

[21] B. D. Greenshilds, "A study in highway capacity," *Highway Res. Board Proc*, vol. 14, pp. 468-477, 1934.

[22] J. P. Hespanha, D. Liberzon, and A. S. Morse, "Hysteresis-based switching algorithms for supervisory control of uncertain systems," *Automatica*, vol. 39, pp. 263-272, 2003.

[23] P. Hingwe, J. Wang, M. Tai, and M. Tomizuka, "Lateral control of heavy duty vehicles for automated highway system: Experimental study on a tractor semi-trailer," California PATH Program UCB-ITS-PWP-2000-01, 2002.

[24]   P. A. Ioannou, *Automated Highway Systems*: Plenum, 1997.

[25]   P. A. Ioannou and A. Datta, "Robust Adaptive-Control - a Unified Approach," *Proceedings of the IEEE*, vol. 79, pp. 1736-1768, 1991.

[26]   P. A. Ioannou and M. Stefanovic, "Evaluation of ACC vehicles in mixed traffic: lane change effects and sensitivity analysis," *IEEE Transactions on Intelligent Transportation Systems*, vol. 6, pp. 79-89, 2005.

[27]   P. A. Ioannou and J. Sun, *Robust Adaptive Control*: Prentice Hall, 1996.

[28]   P. A. Ioannou and T. Xu, "Throttle and brake control systems for automatic vehicle following," *IVHS Journal*, vol. 1, pp. 345-377, 1994.

[29]   S. C. Jolibois and A. Kanafani, "An assessment of IVHS-APTS technology impacts on energy consumption and vehicle emissions of transit bus fleets," California PATH Research Report UCB-ITS-PRR-94-19, 1994.

[30]   W. D. Jones, "Building safer cars," in *IEEE Spectrum*, vol. 39, 2002, pp. 82-85.

[31]   M. Jun and M. G. Safonov, "Automatic PID tuning: an application of unfalsified control," *Proceedings of the 1999 IEEE International Symposium on Computer Aided Control System Design*, pp. 328-333, 1999.

[32]   R. E. Kalman, "Design of a self optimizing control system," *Transactions of the ASME*, vol. 80, pp. 468-478, 1958.

[33]   T. Kelly, "Keeping trucks on track," in *ITS World*, July/August 1999, pp. 20-21.

[34]   P. Kokotovic and M. Arcak, "Constructive nonlinear control: a historical perspective," *Automatica*, vol. 37, pp. 637-662, 2001.

[35]   E. B. Kosmatopoulos and P. A. Ioannou, "A switching adaptive controller for feedback linearizable systems," *IEEE Transactions on Automatic Control*, vol. 44, pp. 742-750, 1999.

[36]   E. B. Kosmatopoulos and P. A. Ioannou, "Robust switching adaptive control of multi-input nonlinear systems," *IEEE Transactions on Automatic Control*, vol. 47, pp. 610-624, 2002.

[37]   E. Kozan, "Optimising container transfers at multimodal terminals," *Mathematical and Computer Modelling*, vol. 31, pp. 235-243, 2000.

[38]   G. Kreisselmeier, "An Indirect Adaptive Controller with a Self-Excitation Capability," *IEEE Transactions on Automatic Control*, vol. 34, pp. 524-528, 1989.

[39]   M. Krichman, E. D. Sontag, and Y. Wang, "Input-output-to-state stability," *Siam Journal on Control and Optimization*, vol. 39, pp. 1874-1928, 2001.

[40]   M. Krstic, I. Kanellakopoulos, and P. Kokotovic, *Nonlinear and Adaptive Control Design*: John Wiley & Sons, Inc, 1995.

[41]   D. Liberzon, A. S. Morse, and E. D. Sontag, "Output-input stability and minimum-phase nonlinear systems," *IEEE Transactions on Automatic Control*, vol. 47, pp. 422-436, 2002.

[42]   C.-I. Liu, H. Jula, and P. A. Ioannou, "Design, simulation, and evaluation of automated container terminals," *IEEE Transactions on Intelligent Transportation Systems*, vol. 3, pp. 12-26, 2002.

[43]   X. Y. Lu and J. K. Hedrick, "Practical string stability," presented at 18'th IAVSD Symposium on Dynamics of Vehicles on Roads and Tracks, Atsugi, Kanagawa, Japan, 2003.

[44]   B. Martensson, "The Order of Any Stabilizing Regulator Is Sufficient a Priori Information for Adaptive Stabilization," *Systems & Control Letters*, vol. 6, pp. 87-91, 1985.

[45]   A. D. Mason and A. W. Woods, "Car-following model of multispecies systems of road traffic," *Physical Review E*, vol. 55, pp. 2203-2214, 1997.

[46]   D. Mcfarlane and K. Glover, "Robust control design using normalized coprime factor plant description," in *Lecture Notes in Control and Information Sciences*: Spring-Verlag, Berlin, 1990.

[47]   D. McRuer, I. Ashkenas, and D. Graham, *Aircraft Dynamics and Automatic Control*: Princeton University Press, Princeton, New Jersey, 1973.

[48]   R. H. Middleton, G. C. Goodwin, D. J. Hill, and D. Q. Mayne, "Design issues in adaptive control," *IEEE Transactions on Automatic Control*, vol. 33, pp. 50-58, 1988.

[49]   D. E. Miller and E. J. Davison, "An adaptive controller which provides an arbitrarily good transient and steady-state response," *IEEE Transactions on Automatic Control*, vol. 36, pp. 68-81, 1991.

[50]   A. Morse, "Global stability of parameter-adaptive control systems," *IEEE Transactions on Automatic Control*, vol. 25, pp. 433-439, 1980.

[51]   A. S. Morse, "Towards a unified theory of parameter adaptive control: tunability," *IEEE Transactions on Automatic Control*, vol. 35, pp. 1002-1012, 1990.

[52] A. S. Morse, "Supervisory control of families of linear set-point controllers Part I. Exact matching," *IEEE Transactions on Automatic Control*, vol. 41, pp. 1413-1431, 1996.

[53] A. S. Morse, "Supervisory control of families of linear set-point controllers .2. Robustness," *IEEE Transactions on Automatic Control*, vol. 42, pp. 1500-1515, 1997.

[54] A. S. Morse, D. Q. Mayne, and G. C. Goodwin, "Applications of hysteresis switching in parameter adaptive control," *IEEE Transactions on Automatic Control*, vol. 37, pp. 1343-1354, 1992.

[55] E. Mosca and T. Agnoloni, "Inference of candidate loop performance and data filtering for switching supervisory control," *Automatica*, vol. 37, pp. 527-534, 2001.

[56] T. Murata, "Petri nets: Properties, analysis and applications," *Proceedings of the IEEE*, vol. 77, pp. 541-580, 1989.

[57] K. S. Narendra and J. Balakrishnan, "Improving transient-response of adaptive-control systems using multiple models and switching," *IEEE Transactions on Automatic Control*, vol. 39, pp. 1861-1866, 1994.

[58] K. S. Narendra and J. Balakrishnan, "Adaptive control using multiple models," *IEEE Transactions on Automatic Control*, vol. 42, pp. 171-187, 1997.

[59] K. S. Narendra, J. Balakrishnan, and M. K. Ciliz, "Adaptation and Learning Using Multiple Models, Switching, and Tuning," *IEEE Control Systems Magazine*, vol. 15, pp. 37-51, 1995.

[60] K. S. Narendra, Y. H. Lin, and L. S. Valavani, "Stable adaptive controller-design .2. proof of stability," *IEEE Transactions on Automatic Control*, vol. 25, pp. 440-448, 1980.

[61] K. S. Narendra and R. V. Monopoli, *Applications of Adaptive Control*. New York: Academic Press, 1980.

[62] L. A. Pipes, "An operational analysis of traffic dynamics," *Journal of Applied Physics*, vol. 24, pp. 271-281, 1953.

[63] B. Richard, "Japan's demo 2000 wows attendees," in *ITS World*, January/February 2001, pp. 18-19.

[64] B. Richard, "IVI's time is now," in *ITS World*, March/April 2001, pp. 10-11.

[65]   M. G. Safonov and T.-C. Tsao, "The unfalsified control concept and learning," *IEEE Transactions on Automatic Control*, vol. 42, pp. 843-847, 1997.

[66]   K. Santhanakrishnan and R. Rajamani, "On spacing policies for highway vehicle automation," *IEEE Transactions on Intelligent Transportation Systems*, vol. 4, pp. 198-204, 2003.

[67]   G. Scora, H. C. Zhou, T. Younglove, and M. J. Barth, "Development and calibration of a heavy-duty diesel modal emissions model," *12th Coordinating Research Council On-Road Emissions Workshop*, 2002.

[68]   R. Scrase, "Driving freight forward," in *ITS International*, May/June 1998.

[69]   S. Sheikholeslam and C. A. Desoer, "Longitudinal control of a platoon of vehicles with no communication of lead vehicle information: a system level study," *IEEE Transactions on Vehicular Technology*, vol. 42, pp. 546-554, 1993.

[70]   S. E. Shladover, "Review of the State of Development of Advanced Vehicle Control-Systems (Avcs)," *Vehicle System Dynamics*, vol. 24, pp. 551-595, 1995.

[71]   J.-J. E. Slotine and W. Li, *Applied Nonlinear Control*. Upper Saddle River, NJ: Prentice-Hall, 1991.

[72]   E. D. Sontag and Y. Wang, "Notions of input to output stability," *Systems & Control Letters*, vol. 38, pp. 235-248, 1999.

[73]   M. Stefanovic, "Safe Switching Adaptive Control: Stability and Convergence (Ph.D. Dissertation)," University of Southern California, 2005.

[74]   M. Stefanovic, A. Paul, and M. G. Safonov, "Safe adaptive switching through infinite controller set: Stability and convergence," *Proceedings of the 16th IFAC World Congress*, 2005.

[75]   G. Stein, "Adaptive flight control: A pragmatic view," in *Applications of Adaptive Control*, K. S. Narendra and R. V. Monopoli, Eds.: Academic Press, New York, 1980.

[76]   D. Swaroop and J. K. Hedrick, "String stability of interconnected systems," *IEEE Transactions on Automatic Control*, vol. 41, pp. 349-357, 1996.

[77]   D. Swaroop and K. R. Rajagopal, "Intelligent cruise control systems and traffic flow stability," *Transportation Research Part C-Emerging Technologies*, vol. 7, pp. 329-352, 1999.

[78]   D. Swaroop and K. R. Rajagopal, "A review of constant time headway policy for automatic vehicle following," *Proceedings of 2001 IEEE Intelligent Transportation Systems*, pp. 65-69, 2001.

[79]   Y. Tan and I. KANELLAKOPOULOS, "Longitudinal control of commercial heavy vehicles: Experiment implementation.," California PATH Program UCB-ITS-PRR-2002-25, 2002.

[80]   C. Thorpe, T. Jochem, and D. Pomerleau, "The 1997 automated highway free agent demonstration," *IEEE Conference on Intelligent Transportation System*, pp. 496-501, 1997.

[81]   K. S. Tsakalis and P. A. Ioannou, *Linear Time Varying Plants: Control and Adaptation*: Prentice-Hall, 1993.

[82]   I. F. A. Vis, "Survey of research in the design and control of automated guided vehicle systems," *European Journal of Operational Research*, vol. 170, pp. 677-709, 2006.

[83]   J. Wang and R. Rajamani, "Adaptive cruise control system design and its impact on highway traffic flow," *Proceedings of the 2002 American Control Conference*, vol. 5, pp. 3690-3695 vol.5, 2002.

[84]   R. Wang, A. Paul, M. Stefanovic, and M. G. Safonov, "Cost-detectability and stability of adaptive control systems," *Proceedings of the 44th IEEE Conference on Decision and Control*, 2005.

[85]   R. Wang and M. G. Safonov, "Stability of unfalsified adaptive control using multiple controllers," *Proceedings of the 2005 American Control Conference*, pp. 3162-3167 vol. 5, 2005.

[86]   H. P. Whitaker, J. Yamron, and A. Kezer, "Design of model reference adaptive control systems for aircraft," Instrumentation Laboratory, M.I.T. Press, Cambridge, Massachusetts, Report R-164, 1958.

[87]   J. C. Willems, "Mechanisms for Stability and Instability in Feedback-Systems," *Proceedings of the IEEE*, vol. 64, pp. 24-35, 1976.

[88]   H. J. Xu and A. Ioannou, "Robust adaptive control for a class of MIMO nonlinear systems with guaranteed error bounds," *IEEE Transactions on Automatic Control*, vol. 48, pp. 728-742, 2003.

[89]   D. Yanakiev and I. Kanellakopoulos, "Analysis, design, and evaluation of AVCS for heavy-duty vehicles: Phase 1 report," California PATH Research Report UCB-ITS-PWP-95-12, 1995.

[90]   D. Yanakiev and I. Kanellakopoulos, "Engine and transmission modeling for heavy-duty vehicles," California PATH Research Report Tech Note 95-06, 1995.

[91]   D. Yanakiev and I. Kanellakopoulos, "Nonlinear spacing policies for automated heavy-duty vehicles," *IEEE Transactions on Vehicular Technology*, vol. 47, pp. 1365-1377, 1998.

[92]   D. Yanakiev and I. Kanellakopoulos, "Longitudinal control of automated CHVs with significant actuator delays," *IEEE Transactions on Vehicular Technology*, vol. 50, pp. 1289-1297, 2001.

[93]   K. Yi, S. Lee, and Y. D. Kwon, "An investigation of intelligent cruise control laws for passenger vehicles," *Journal of Automobile Engineering, Proceedings of the Institution of Mechanical Engineers Part D*, pp. 159-169, 2001.

[94]   J. Zhang and P. A. Ioannou, "Control of heavy-duty trucks: Environmental and fuel economy considerations," California PATH Research Report UCB-ITS-PRR-2004-15, 2004.

[95]   J. Zhang and P. A. Ioannou, "Longitudinal control of heavy trucks in mixed traffic: Environmental and fuel economy considerations," *IEEE Transactions on Intelligent Transportation Systems*, 2006.

[96]   J. Zhang, P. A. Ioannou, and A. Chassiakos, "Automated container transport system between inland port and terminals," METRANS 2002.

[97]   W. B. Zhang, "National automated highway system demonstration: A platoon system," *IEEE Conference on Intelligent Transportation System*, 1997.